A WEASEL
IN MY MEATSAFE

Also by Phil Drabble

A WEASEL
IN MY MEATSAFE

Phil Drabble

with drawings by Ralph Thompson

MICHAEL JOSEPH

LONDON

Originally published by William Collins Ltd, 1957
This revised edition first published
in Great Britain by Michael Joseph Ltd
52 Bedford Square, London WC1
1977

ISBN 0 7181 1635 6

Printed in Great Britain by
Hollen Street Press Ltd, Slough,
and bound by Redwood Burn, Esher

CONTENTS

PREFACE

I was born on the edge of a Staffordshire town that was about the most unlikely habitat to produce a naturalist. My father, who was a Black Country doctor, had neither the time nor inclination for country pursuits but his mining practice included a number of farms and, luckily for me, many of his patients became our family friends.

As a child, these farmers allowed me to run wild on their land and I grew up with more broad acres to roam over than if I had been heir to a stately home. My earliest memories are of catching newts and dragonflies from the 'swags' or subsidence pools at the foot of worked-out pits.

I had a natural affinity with 'bad' characters so I learned more natural history from the poachers and keepers and rat-catchers who were my father's patients than from boffins and desiccated school gaffers.

Because I was unable to spend as long in the wild places as I would have liked, I brought as much of the country-side as possible home with me. I kept caterpillars in boot boxes, newts and sticklebacks in jam jars, a rat on the verandah and even a weasel in my meatsafe!

I had a ferret and a dog by the time I was ten and, at twelve, I knew far more than was respectable about the ancient art of poaching.

Although I was pitched into a factory to earn my living, and it took me twenty-odd years to escape, I was immensely lucky in the friends I made.

A Weasel in my Meatsafe

The late Miss Frances was one of the best naturalists I ever met. She was a distinguished writer and nature photographer who combined being Master of hounds with her love of conventional natural history. She had reared and kept an incredible variety of British birds and mammals and was always generous with her experience when I wished to follow suit. She seemed to know everyone in the wildlife world and used to invite my wife and me to her famous luncheon parties to meet them on level terms. I learned more in a day's visit to her house near Bridgnorth than I would have done in a term at university.

Brian Vesey-FitzGerald, first as editor of *The Field* and later as an editor publishing books, encouraged me to start writing and broadcasting. Without his help, I might still be languishing in a factory.

There must be scores of town boys like me who are fired by an overwhelming urge to escape to wilder places and saner values. I have never had a period in my life that I would swap for the period before, and I am working harder now and am happier than I have ever been. I wish all youngsters who share my tastes a similar slice of luck.

1977   P.D.

CHAPTER ONE

CHILDHOOD

My early memories are punctuated by the rollicking giggles of my nursemaid. Whenever anything pleased her – and she was usually pleased – she gave vent to belly-shaking brays that were just as loud and infectious when she gasped for breath as when her mirth was in its swelling prime.

'Nursemaid', of course, is mere exaggeration. She was a local girl from across the street, who came in the afternoons to take me for a walk between lunch and tea. Her name was Elsie.

I remember her with the greatest affection for all the fun we had. Our walks were necessary because the house I lived in was in the very centre of a Black Country town. It was semi-detached with a grey flagged yard at the back, that streamed a dismal sooty soup when it was wet, and sizzled when it was dry and hot.

The trams from Walsall passed five yards in front of the dining-room window and, behind the trams, a lane wandered down first between more rows of grimy brick houses and then through fields. Cemetery Lane it was called and sometimes we went that way for a walk to gape, in morbid wonder, at the shiny black horses and flowers and brassy coffins and mourners. But if it got dusk while we were away, it was necessary to come back home all the way round by High Street, because neither Elsie nor I had much stomach for cemeteries after dark.

It was pleasanter, therefore, to go the other way. Our yard opened into another, much bigger courtyard, with stables and coachhouse and all the trappings of Victorian respectability.

Our landlord, from the other half of our semi-detached, kept a cart-horse or two here for his trade as coal merchant, and a hunter for his pleasure.

Here then, right in the heart of the industrial Midlands, was stabling for hunters. Hunters which could walk from their loose-boxes straight into a green field edged by tram-lines to the east, houses to the north and south and, miraculously, more green fields to the west.

This, of course, was the way Elsie and I used to go when we weren't watching funerals. Two fields away was a blast furnace where the chap who worked the shunting engine sent us both into paroxysms of mirth with his wit. And when my sense of humour, which was less highly developed than theirs, became surfeited with the repetition, I would wander off and dabble in the 'swags' or pools caused by mining subsidence.

There are lots of these pools in the district, varying from a few yards across to many acres. They hold pike and roach and perch for the grown-ups, and newts and sticklebacks by the score for the kids.

Elsie was a wonderful escort for these expeditions. I was an only child, left very much to make my own amusements. She, on the other hand, had a favourite brother who played goal for the church football team. In his spare time he made me kites, taught me to spin tops and, best of all, made fishing-nets out of the cotton bags that held his pigeon grit.

We caught newts in these by two methods. One, which I regard as the hamfisted way, was to draw the net firmly through the weed, the elodea and rushes that choked the edge, until it was full and heavy with weeds and mud. Then, like monkeys picking fleas, we would sit on our haunches on the bank, gently picking over the foetid mass, and pop anything that moved into a jam jar.

There were beetles and water scorpions and water spiders and dragonfly larvae, for none of which we knew the proper name. Occasionally the whole muddy bag would heave and tremble, and Elsie's breath would come in short gasps of excitement which would eventually break surface in a crescendo of guffaws.

There was only one possible explanation. We'd got, in our bag of slime, another newt. 'Askers' she called them and she was so certain they spat fire that I marvel, in retrospect, at her courage in picking them up at all. But once in our net they had no chance. They promptly joined the motley crew in the jam jar.

We developed a more scientific and amusing way of catching them, though. We discovered that every few minutes a newt would rise to the surface, blow a bubble and sink again. A little dexterity was all that was necessary to pop a net in his path, before he submerged, and his capture was certain.

My mother and father, who did not share my early passion for catching things, were surprisingly tolerant. When Elsie and I returned in triumph with some new batch of captives, they whipped up interest and gave us jam jars to put them in. By morning the fish had usually died and the newts and beetles escaped. I was blissfully ignorant that fish would suffocate if I put them too thick in a jar, that water beetles could fly or newts could climb. And in my ignorance I did cruel things which make me shudder to remember. My very favourite newt Tibby, for instance, was left out on the window-sill one bitter night with only a pound jam jar of water for protection. Next morning, like a trout in aspic, she lay stiff in the centre of a block of solid ice. My experiences down Cemetery Lane were not wasted. I sobbed in mourning, buried Tibby in reverence, sprinkled earth on the spot and mumbled platitudes like a true parson.

As usual, it was Elsie who came to the rescue. She presented me with a large goldfish bowl for my birthday, and spent the previous evening catching askers to put in it. Never before had there been such a success. She arrived at the house with the bowl and no less than twenty great-crested newts, varying in size from monsters of five inches to titches which were tadpoles last year.

My mother was overwhelmed. The bowl was spherical, so the newts would be safe; it was more respectable than jam jars so it could come in the house. She made an error of judgement, though. Newts can climb out of a goldfish bowl. When I came down in the morning, like a soldier inspecting his spoils, I found a gleaming new bowl full of crystal-clear water. There wasn't a newt in sight. Mother was horrified. Twenty newts lost in the house. Breakfast forgotten, we were organized into a search-party. Rugs were carried out, furniture moved and carpets upturned without success. We did, I think, find a couple under the stair carpet, and for the next few weeks odd, pathetic mummified corpses were poked from under the skirting-boards and similar cracks where my birthday presents had crept for shelter and stayed to die. But the vast majority simply disappeared without trace.

I took it very hard. At seven years old I was suffering all the pangs of a townbred lad who smelt the smell of wild things for the first time. I lacked the advice and help of any knowledgeable grown-up and, at that time, I hadn't even any children's books on natural history. Elsie was keen to help but it was very much a case of the blind leading the blind. What knowledge she had was so diluted with superstition that I had to spend subsequent years unlearning much that we found out.

She never quite got over her fear of picking up newts, for example. Not because she disliked the feel of 'crawlies' but because she really believed them to be capable of 'spitting fire'. It was a fear that she certainly had never read about, for her tastes did not encompass that sort of reading. It was a belief passed down to her by word of mouth as the creeds of primitive folk traditionally are. It was a tradition hallowed by antiquity, for had not the dragon – the biggest newt of all – spat fire at St George? And the salamander, a second cousin of the newt, is associated with fire by folk-lore down the ages.

So Elsie and I groped about between the facts and fancies of natural history. On the one hand we were deceived into believing that dragonflies were really 'horse-stingers'. On the other we learnt a lot by first-hand observation.

The newts that we caught – and lost – played a most important part. Their colouring was brilliant, their movement in water sinuous and seductive, their habits arresting.

Having upset the house by the loss of one score of reptiles it was simply vital to strike again quickly whilst the iron was hot. To disappoint a child on his birthday would have been unthinkably mean; at any other time to forbid newts in the house would have passed for sound discipline.

By tea-time my new goldfish bowl was stocked once again. A piece of muslin was securely tied over the mouth of the bowl which was put in the place of honour in the centre of the breakfast table.

No one quite knew what newts like to eat. We tried bread and biscuit crumbs and hard-boiled egg. Our newts got as thin as greyhounds. It was Elsie's brother who came to the rescue. He knew that newts ate worms and he sent a supply of nice small ones to prove it.

Looking back, our ignorance seems quite incredible. Indeed, with biology classes in schools, there can't be a child in the country today who wouldn't know. To us the discovery was fascinating. I vividly remember sitting goggle-eyed at breakfast, my porridge lying like frogspawn untouched before me, watching my birthday presents tuck in. They grappled and wrestled for wriggling worms, and a newt eating a worm a little too large is hardly a sight for the breakfast table. It was as if the worm realized the effect of his host's gastric juices, so it writhed in knots to avoid being swallowed. But hungry newts are tough and strong. In the end the best of worms gave in and disappeared. I heaved a sigh of relief that the struggle was over and turned to my tepid porridge.

By the time the eggs came on, someone noticed the largest newt appeared to have sort of palpitations of his tummy. Obviously he had very bad indigestion. That was putting it mildly for the largest worm, the one all the real fuss had been about, popped up again, to all appearances as good as new. The whole grisly pantomime was re-enacted through such a series of rehearsals and performances as nearly nipped my career as a naturalist neatly in the bud.

Not that I minded. I was childishly sympathetic to the worms, but my thirst for knowledge easily outpaced any squeamishness. For the rest of the family, breakfast was untouched and unthinkable. In future, I compromised by never feeding my pets at table.

We found the more intimate habits of newts far more attractive. My birthday is in May and the male newts we had caught were in their most resplendent colours and the females were ripe with spawn.

One evening I sat eating my supper and trying to spin out the time before bed. Suddenly I noticed a female crested newt in what I took to be the throes of constipation. A male had been writhing and cavorting before her with all the sinuous grace of a ballet dancer. Her reply was to expel what appeared to be a tiny string of sausages. I called my father to prescribe a dose for her. And then we noticed the string get longer and each little sausage, as it came in contact with the water, swelled visibly and grew. It became obvious that she was spawning and as her eggs grew longer she reached down and grasped the end one between feet as delicate and sensitive as artists' hands. Each egg that she grasped, she broke off, placed with reverence in the leaves of water plants, wrapped in a tiny parcel and hid it carefully from view. The males, we discovered, were just as fond of eating spawn as fertilizing it. This first newt that we watched laid more eggs than there were water plant leaves, and when there was no further cover to hide her treasures she almost went frantic. There was nothing I could do to help her because the nearest pool was half a mile or so away and I wasn't allowed out after supper. So, more as a joke than anything, I tore up a few scraps of newspaper about as big as confetti and dropped them in the bowl. At once my newt grabbed an egg and wrapped it in newspaper as willingly as she would have used a leaf. I was as excited as an explorer discovering new lands. The rest of the family was mildly intrigued enough to allow the craze (as they thought my new-found joy to be) to wear off.

My first and major battle was won. I had discovered an interest that absorbs me far more now than it did then, and the family were passive in their delight if not active in their help.

How much I owe to those early newts, though, I can never calculate. Even today my breakfast table lacks something without them, though it was they, poor things, which suffered while I learnt. In any doctor's house, for example, hygiene must of necessity rank high. So every morning my goldfish bowl was emptied as regularly as chamber pots in a hospital. Every morning my poor newts were tipped writhing with indignation into the sluice while their home was rinsed and refilled with fresh, sparkling chlorinated tap water, as pure and uninteresting as only tap water can be. They never had the chance of swimming in good mature pond water rich with the microscopic life they love, the temperature changed suddenly from their warm room to cold tap and yet, despite mishandling, their toughness triumphed and they survived.

Not only did they survive. They grew. Which reminds me of the other shock I had. It had got such a habit with me to linger over my meals, chewing my mental cud, that I often sat on alone when the others had gone. A great, brilliant spotted male had, for some days, scooped the lion's share of the worms I put in. So much so that he was visibly growing bigger and fatter. I sat watching him this night, wondering how big he would grow, when right in front of my nose, his skin took a visible longitudinal split.

I was horrified. I'd often heard the local imprecation 'Better bally bost than good things spoil', and I'd often laughed at it. Now I suddenly remembered that the last time I'd pulled rude faces at Elsie, she told me that if the wind changed I'd stick that way. Naturally I took it, as it was meant, as a joke. When I saw the skin split up my newt I began to wonder. I was still quite young enough to remember fairy stories about frogs which turned into princesses. I sat watching this split skin against my better judgement. I wanted to see what he turned into, yet I was terrified lest my worst fears should come to pass.

They didn't. Nothing so romantic. Instead I sat spellbound, went away and came back to watch again as the skin was gradually sloughed off until, eventually, it hung complete in the water like a tiny gossamer ghost. Its owner was left resplendent in a brilliant new coat, big enough to allow for a little more expansion. It swam after its old skin, drifting like a shadow in front of it, rolled it neatly up and ate it. Nature is so lavish in some things yet, in others, too miserly to waste an old, disused skin.

The seeds of that early childhood experience germinated and have blossomed spasmodically ever since. Once I had acquired the taste for an aquarium I repeated the dose again and again. At times, like all children, I had goldfish which ogled unnaturally, for a while, in their little crystal cage, but soon succumbed to the family craze for hygiene and fresh water.

More interesting, I found, were the things we caught ourselves. Little black whirligig beetles which played endless games of tick on the surface and then dived in a vain search for vegetation to hide in. Children are cruel, I think, through sheer ignorance and lack of imagination. At that time I had never seen a proper aquarium where plants and animal life were poised in that miraculous natural balance which keeps both of them healthy. I had never witnessed the underwater vistas of brilliant greens and delicate browns; of movement and loves and underwater battles to stir the heart of an adventurer. The pinnacle of my joy was the pathetic little glass sphere where my prisoners swam in misery and I deluded myself that I was their friend.

Elsie and I, in our wanderings, became intimate with the wilderness that the spent coal-mines and iron works had left in the Black Country. Just as wild hares acquire a territory which they know very well because they travel round and round within its limits, so did we come to know the square mile which is to the immediate south-west of Bloxwich. About all the sterile clay spoil-banks would support, as their contribution to spring, were stunted little coltsfoot flowers. The lime-starved swags gave kingcups and lady-smocks in their season; we picked buttercups and dog-roses; at any season we could usually find golden gorse flowers for, when gorse is not in bloom, then kissing's out of fashion. All the commonest, least romantic of flowers, we picked with the reverence and wonder of botanists making a discovery and my mother put in the place of honour, like a girl with her first rose.

We wandered on Bentley Common, taking elaborate precautions never to get within half a mile of the gypsy caravans, for Elsie had a fear of gypsies as healthy (or unhealthy) as her terror of horse-stingers. To back it up she hinted and implied such wickedness as made me cower as a child, but only until my sense of curiosity was more fully developed than my fear. Years later, when I first went to London, I was similarly warned to keep away from a certain theatre where the naked ladies would corrupt my morals. Elsie's warning about the gypsies had exactly the same effect. Only the gypsies, of course, taught me much more.

Our long detours to avoid them took us over mounds and fields uncrossed by paths. Here and there we would come across some unsuspected pool, sometimes quite large, and at others but a few feet across. Then later, when we were in search of fresh captives, we would go back with our fishing-nets. And there we made a momentous discovery.

In apparently similar pools we caught entirely different animals.

One, grey, forbidding, cindery swag was famous for enormous yellow-bellied black newts and their flamboyant males; another had water-boatmen which swam the right way up, and another contained what looked like the same beetles, except that they swam upside down. In odd pools and at odd times, of course, we got variety. One of our favourite haunts contained more Great Diving beetles *Dytiscus*, and their larvae than anywhere for its size I've ever seen. These viciously destructive great brutes made carnage if put into mixed company in an aquarium. Even amongst themselves they fought such duels and battles that fired my imagination and swamped the kindness of my heart in a surge of elemental sadism. They made a boxing booth look tame, a cock fight civilized.

It would, of course, have been disastrous to have been brought up without male companionship. My landlord's groom kept open house in his stable. My prime joy, I think, was to watch him dealing with Bella. She didn't belie her name. Pure-bred racehorse, Cook said she was, though she could kick and bite like a mule. My interest lay chiefly in her deformity, though. She had a broken wind, to cure which a brass pipe, with a large flange on it, had been stuck through her neck down into her windpipe. Every day Cook used to insert a prehensile finger, draw out this brass pipe and polish it. Meanwhile I gaped agog at the flap of skin closing over the hole convincing me she'd die of suffocation. And when, polished and gleaming, her brassy windpipe was made good, she'd nuzzle her groom as softly as a lover.

The opposite side of the stable to the loose-boxes was the

saddle-room. A wonderfully perfumed sanctum of gleaming leather and steel, with always a fire burning in the little black-leaded grate. On one side of the hob was the old black cat, while on the other, brewed bran mash, saddle soap, or Cook's breakfast according to time and season. Here I learnt a lot of natural history. While he boned his leathers or polished his bits, he would tell me of rats in the trenches or great bats like vampires. On one memorable occasion he produced for me an old cigar box with five squirming young mice, still naked and blind. At that time I had no idea how to deal with them. Cook had found them between some bags of corn and all my life, it seems, people have brought such finds to me.

Anyhow there was no great welcome for my mice at home. I was told in no uncertain terms that they stank, that I should die of typhoid or diphtheria or worse if I touched them, and that it was very cruel to take such infants from their mothers. So back they went to Cook in the stable. I can't think it accidental that the old saddle-room cat took them in charge.

My next excursion into keeping British mammals was even more abortive. I was crossing the yard one day when I saw a young rat about the size of a man's thumb. It limped from some injury and was even slower and more clumsy in its movements than I was. Without a moment's thought I stooped to pick it up. Far quicker I yelped with pain and tried to put it down again. That wasn't so easy. He'd sunk his wicked yellow teeth into my right thumb right up to the gums. I picked up my hand and he hung there, for a moment, like a ferret, before dropping off to amble safely away. Like a young huntsman I had been blooded for the first time.

Looking at that thumb now there are so many scars that I find it impossible to know if I still bear the marks of that first rat or if the trademark he left has been obliterated by his descendants. But the first finger and thumb on my right hand have since been mauled so often, in the cause of animal friendships, that sometimes I lie in my bath, scrub until the white tissue stands out, and decipher from my mutilations the pedigrees of the successors of my opponent in that far-off stable yard.

CHAPTER TWO

SCHOOL

I bear a scar to this day which reminds me of one of the few episodes I didn't hate at school. On the whole, I hated the place, my masters and my schoolfellows. I felt caged and cooped up, obliged to subsist on their plain food, for one ate that or went without; obliged to live in a ruck like rabbits, when I ached for the solitude of hares; obliged to wax enthusiastic over team games, when I longed to wander by myself in the countryside.

Miss Lonsdale, my matron, understood my frustration, though there was little she could do about it. Even when she was ill she thought of others.

She had been gravely ill in the school sanatorium and, for some reason I have never understood, they had used medicinal leeches on her. Two of these leeches were confined in a bottle in her sickroom and, as soon as she felt a little better, she sent someone to 'find out if Drabble would like them'.

I was naturally quite delighted and not a little overwhelmed. Years ago, in Elsie's day, I had kept common horse-leeches in my aquarium and been enchanted by their unexpected sinuous grace when they swam. They crawled in a series of slow-motion loops, like caterpillars, and then quite suddenly swam gracefully off like butterflies taking wing.

Superficially, Miss Lonsdale's leeches looked very like the ones I'd caught in the swags and pools of my childhood. Olive green like long narrow snails, very pointed at the 'head' end.

All agog with excitement, I asked my biology master if I might keep them in his lab. He was widely known as 'Bunny' and I imagine, since I never did much work, that he liked me about as well as most rabbits like most stoats. But since Miss Lonsdale had sent them from her sick-bed, a first wish on recovery instead of a last wish on departure, it would have been too ungracious to refuse to give them sanctuary.

So next day they were swimming round a newly cleaned aquarium, gently probing every crevice with their trunk-like fore-ends for a possibility of escape. As soon as the next biology class assembled I enquired of my mentor what they should feed on. 'Blood,' he barked with most unrabbit-like ferocity. So I caught one of the frogs he had purchased for us to dissect and held it gently but firmly cupped in the palms of my hands. Now, in common with most cold-blooded creatures – which aren't so much 'cold-blooded' as controlled by the atmospheric temperature – frogs do not tolerate any sudden great change of temperature. My hands were much warmer than the frog, which quickly died. So I popped the corpse, untainted by anaesthetics, straight into my tank of leeches. They took not the slightest notice.

Next morning I was up early with my catapult and managed to get close enough to a young sparrow to knock him off his perch. I didn't waste any time at all. I plucked a bare patch of breast and popped him into the tank. Still not a sign of interest. Indeed, whatever I popped in was treated with the same disdain and each day my beloved leeches shrank just a tiny bit more. I began to visualize Miss Lonsdale, when she was completely recovered, coming to see the leeches, which had saved her life, and finding nothing but two withered scraps of skin as memorial to a fashion that has died.

By next biology lesson I was desperate. 'What *do* you feed these things on?' I demanded. 'Blood,' roared my teacher. 'I've tried that,' I said. 'What sort of blood?' 'Yours,' spat old Bunny, with the sort of gleam in his eye that masters get when they're going to beat you. He thought that would certainly shut me up. He reckoned without my regard for Miss Lonsdale.

So I fished out my leeches, put them on the palm of my hand and sat back to watch developments. So did the rest of the class. Watching me offer myself as a human sacrifice was better than all the biology classes in the session.

But the leeches didn't seem interested. One refused to move at all and sulked, while the other crawled off my hand up my arm on a tour of inspection. He didn't get very far. In the centre of my forearm, on the soft inside, the same side as my palm, he stopped. Quite deliberately he began to bore his way in and, when he'd got a good hold, I held up my arm so that he could dangle in full view of the whole class. They were enthralled and gathered round agape. As the blood began to flow, it pulsated down my leech until it seemed as if he was being milked by some ghostly hand. More and more blood flowed out of me and into him and his bottom swelled until he looked exactly like a tiny Indian club. My audience was thrilled, and even I quite enjoyed it because it hurt surprisingly little. About like a nettle sting.

At last Bunny began to get uneasy. Even he didn't want me to die of loss of blood in his class. So he told me to get it off.

That was easier said than done. I gently caught hold of its tail and pulled. It stretched like catapult elastic, but there was no sign of it letting go. Bunny thought I was putting on an act for the benefit of the form, and roared at me to get my assailant off. I asked him how. At last even I tired of the novelty. I picked up a pair of dissecting tweezers and gently attacked the business end, which was by now firmly embedded in the fleshy white part of my forearm. The result was magic. A very bloated leech let go.

The class was requested to start a little work, whilst I was given orders to wash my arm.

Half an hour later I had soaked two handkerchiefs without any sign of the bleeding stopping. Never before or since have I seen so much blood from such a tiny wound. Tea-time came and I was still bleeding, so I reported to the sick-room for attention, only to be passed on to the local doctor at his surgery.

He took one look and asked who the devil had told me to

put a leech on that part of my arm. I referred him to my biology master. He snorted and confirmed an opinion that I had imagined to be only my own biased view.

Apparently leeches inject a substance called hirudin, specifically to prevent blood clotting. So the approved way of applying a medicinal leech is to put him in a test-tube and hold the mouth of the tube over a bony part of the body, such as the temple or the top of the cheek-bone, if it is desired to remove the colouring of a black eye. The leech can then only fasten to the tiny area covered by the mouth of the test-tube and can never puncture anywhere where it will be difficult to apply subsequent pressure to stop the bleeding. And to get him off, you apply a pinch of salt which makes him loose at once. The danger in pulling is that the beast may break, leaving his head attached to fester.

My leech, of course, had had a field day. He had crawled purposefully up my arm to fetch up over the vein that runs from the inside of the elbow to the wrist. He'd then bored straight through the wall of this vein so that the doctor found it necessary to stitch me up to stop the bleeding. The rude message he invited me to pass on to my biology master, telling him the missing facts about the care of leeches, is one of the happiest memories of school and the scar on my arm reminds me of it every time I have a bath.

I began falling foul of Authority over my pets years before at my prep. school. Someone had given me a grass snake which survived, most unnaturally, for a while in a cardboard boot box. I fed it occasionally on small frogs which it rarely ate, and, for exercise, I used to put it in my shirt, where it amused itself by chasing its tail round my tummy, and the class when it poked a wicked-looking head and forked tongue down my sleeve or over the neck of my collar.

Its career with me was cut short, though. One terrible day I wore braces instead of a belt. I heard the master coming down the passage and, just in time, popped my snake out of sight down the front of my shirt. Too late, I realized there was no belt to maintain it securely above my waist. It wriggled down into my trousers, gave me a moment of acute embarrassment, and plopped out on to the floor. I just had time to discover that snakes were neither cold nor slimy before this, the only one I ever kept, was confiscated. But somehow I have never managed to work up any sort of enthusiasm about keeping them and can't really claim I minded.

I didn't really learn much about keeping anything at my prep. school, where pets were never welcome. We were allowed tortoises, which we kept within bounds by boring a small hole in their shells just above the tail, putting a small split ring through it, and tethering them on a long piece of string. I found even more effective than pegging them down, was to let them trail a couple of yards of brilliant tape, which was easy to locate, even if the old tortoise had gone to ground in the dense cover of a bush or clump of flowers. But apart from a passing craze for a mongoose – which I've never had to this day – I was never particularly keen to own anything but British animals or birds or insects.

What I did learn at my prep. school was the thrill of pitting my wits against the birds and animals around me. The school gardener hated coming to feed the fowls and water the school horse on Sunday afternoons while, for me, it was a pleasant change. In return he used to bribe me with fruit from the garden which was out of bounds to the boys.

And then, out of the blue, he showed me how to set horse-hair snares for the sparrows which stole the fowl corn.

I was spellbound. My parents would have been shocked at what they would have regarded as corruption of my youthful innocence. Indeed, each bird I caught gave me qualms which almost made me feel sick. But savagery must lie so close to the veneer of civilization that, in spite of my nausea, I was fascinated. I simply could not resist doing it again.

The process was simple. A loop in the end of a stout horsehair – plucked in peril from the tail of the school horse – was made into a running noose by passing the free end through it. This noose was arranged to be just slightly smaller than a hole in the wire netting to which it was tied. Unwary sparrows thrusting through for food literally put their heads in my nooses and were hanged.

Were it as simple as that there would have been no excitement. The subconscious appetite of the hunter is only whetted by difficulty. I had been brought up with a background of urban tastes, yet the whiff of a quarry had intoxicated me like a foxhound puppy the first time he goes cub-hunting.

I used to steal away quietly down the school garden and dangle my insidious nooses over the wire netting of the fowl pens. Almost by instinct I discovered that the sparrow did not pop through any hole of the netting haphazardly. A noose set over nine holes out of ten would lie undisturbed for days. Set it in the tenth hole and there would be a victim in as many minutes. At first I lay painfully concealed in thorny gooseberry bushes straining to watch which holes the birds most used. Then I noticed that as they alighted on the netting they usually passed their droppings whilst they checked the coast was clear. After that, of course, it was simple to trace the vertical lines of whitened wire and set my snare on the hole above the topmost stain.

The result was extraordinary. Quite suddenly it became easy. Every noose, next day, had a corpse, but instead of each quarry being a monument to my skill, the pathetic bundles of feathers were a tax on my conscience. The thrill of the hunter had evaporated. Now it was only butchers' work. But at least I had learnt that sparrows were as much creatures of habit in their routines as rats or rabbits, whose padded runs are so much more obvious.

This love of catching things grew hand in hand with an insatiable curiosity to learn more about them. As I grew older, I devoured every book I could lay hands on remotely connected with my interests. I read old books on hunting and hawking; books on bird photography, dog training and gamekeeping; books on camping and canoeing; books about anything that could help me escape crowds and get into the country.

At my public school I had a bit of luck. My house-master, E. H. Furness, was the scholarly sort of man least calculated to tolerate a mutinous young savage like me. Indeed, his dress was so immaculate, his speech so precise and his manner so mild that he fell prey to more than his fair share of ragging. We nicknamed him Bobby, because he was so polite he actually asked boys to do things instead of telling them. So he became the Pleaseman – or Bobby! We were an ill-assorted couple with but one common bond. He was the President of the school Natural History Society – 'Bug Soc.' as it was irreverently termed – and he did more to make my life bearable than the rest of the staff together.

I suppose, like other boys, my initial interest in this

'Bug Soc.' had other motives than were apparent on the surface. It held periodic rambles to places like Ran Dan woods and Chaddesley Corbett. The Ran Dans encompass very many acres, and for lads who want to escape for a surreptitious smoke, the cover they afford is as good as it is for foxes and birds. And the teas that were provided at Chaddesley Corbett had only the monotony of school food to compete with. For us, who were even less tolerant of restraint, a farm in the woods used to sell cake and a jug of rough Worcestershire cider for a very moderate bribe. The natural history we learnt on these expeditions was therefore incidental to our main ambition to escape for a while from authority.

My first jackdaw was a shining exception. I was up in a thicket one day, quietly puffing a cigarette without particularly enjoying it, when I noticed an old jackdaw down the path feeding a young one on the ground. Indeed they made so much squawking and pother that I must have heard them had I been almost stone deaf.

At once my heart turned over with the primitive urge to hunt. I had long ago realized that the very first principle of woodcraft is never to make the least sudden, jerky movement. My hand crept automatically to my pocket. Long seconds later it slid out imperceptibly gripping my catapult. I loaded and pulled with bated breath. The stone sizzled across the ride to land 'thock' in the side of the old bird, which rolled over with scarcely a quiver.

As usual, the reaction was sickening. The preamble and blow had been instinctive. When reason returned I realized that I had wantonly created an orphan. There in the clearing before me was a bewildered young bird vainly waiting for its parent to feed it. I picked it up and carried it tenderly back to school.

Bobby was annoyed with me and full of compassion for my jackdaw. The kindest thing, he felt, was to pull its neck and put it out of its misery. But he understood my remorse too. He knew instinctively that if I reared it successfully, as atonement, I would emerge a naturalist at heart; that if he rubbed my nose in the fact that I had slaughtered a nursing mother by sacrificing her chick as well, I should probably join my schoolmates in their herd pursuits.

I was therefore given permission to keep him. A disused carpenter's shop was put at my disposal and young Jack began his sojourn with me in an old packing-case in the shed. Neither my games nor my studies got very high priority in my thoughts at school. After I acquired this jackdaw I thought of nothing else.

The first problem was food. Partly because it was easy to steal from the dining-hall, he started off on bread and milk. It subsequently turned out that I could scarcely have chosen anything better. Although bread doesn't suit all young things, milk usually does, and I have started a whole range of pets from wild duck to weasels on this, the simplest of all diets. I tried him on suet pudding, sago, dumpling and the other stodge which appeared at table. My jackdaw was not impressed. Then I tried hard-boiled egg. It worked like a charm and on a mixture of this with bread and milk he never looked back.

I used to get up before breakfast to feed him and clean out his box. Then I fed him in our ten-minute mid-morning break, at lunch-time, after afternoon school, tea-time and supper-time. He grew like a willow and seemed to regard me in every way as his mother.

Various members of the staff were slightly incredulous. That I, who was habitually late for morning prayers, should willingly be abroad when others were in bed, seemed unnatural; that I should cloister myself alone in 'my' old carpenter's shop without supervision and still without misbehaving, seemed to be quite out of character.

They were right, of course. As soon as the door shut behind me, I lit up a cigarette. Everyone smoked on all Natural History expeditions; it was traditional. And I was well-known for outdoing tradition by keeping an old suit of clothes, quite different from the school uniform, in the school museum of which I was curator. I used to sneak out at night, dress like some lad from the town and swagger out of bounds from sheer bravado.

Small wonder then that the rest of the staff looked at my housemaster as if he was scarcely normal. Even I, who held him in high regard, was slightly doubtful of his motive. I wondered if his concern for my black orphan was genuine or if he was merely loosing out enough rope to hang me.

I needn't have worried. Two or three times a week he paid a ceremonial visit to see for himself how the rearing progressed. He made no sign if he smelt tobacco, though once the atmosphere was thick when he came through the door and I was trying to fan clouds of billowing haze through the window. But it was Jack we both wanted to see and we fell into the jargon of naturalists forgetting, for a while, our normal relationship.

Like a gardener with green fingers, it seemed, I couldn't go wrong. I'd been lucky, of course, to start with a fledgeling so strong and well-grown. His change of diet didn't seem to have the slightest effect and within a week he was flying strongly the full length of his shed.

About this time I was devouring every book I could lay hands on dealing with animals or birds and I remember coming across the article on Falconry in the *Encyclopaedia Britannica*. It is a very good article indeed and fired my ambition to own a hawk. Poor old Jack was the next best thing, and I started to experiment on him.

The first thing you teach a hawk, it seemed, was to feed 'from the fist'. So before my jackdaw got his boiled egg or bread and milk, he had to perch on one hand and feed from the other. That was easy. If he was hungry – and his appetite was insatiable – he'd cheerfully perch anywhere at all to be fed.

The next step, I read, was to get your hawk to 'fly to the lure'. This meant that you got a round piece of baize-covered wood which you swung round your head on a long cord like a ferret line. The hawk sees this lure gyrating from afar, stoops down to it and begins to feed as soon as it comes to rest on the ground. It is then a simple matter to show even more attractive food than is fastened to the lure so that he hops tamely on the fist to be fed.

The first and only time I tried a lure on my jackdaw, I nearly lost him. I made a very good copy of a hawk lure and swung it round my head like a falconer. But Jack did not approve. He panicked and flew into the window in a frenzy. The glass held and he fluttered half-stunned to the floor. I leapt across the room to the rescue, only to make him panic more. He struggled from the floor and flew smack into the far wall. I panicked too. The wilder he got the more I chased him round in a misguided effort to save him.

In the end I was properly flustered and he was panting

and exhausted. I had started with a trusting young bird, which was finger tame, and finished, fifteen seconds later, with a cowering bundle of feathers as wild as his brothers.

So I started all over again. I left him quite alone to recover his composure a little. Four hours later the pangs of hunger overcame his new-found fear enough to force him to risk all for the first few beakfuls. Then he remembered his fright and flapped off like a moth at a candle. I went away and tried next morning, but it was several days before his nerves began to steady.

One ill-considered action on my part had undone in a few seconds the work of weeks. Hours and hours of patience were sacrificed for an instant of thoughtlessness.

For the next week, however I cajoled, my young bird was jumpy and shy. Hunger was but a temporary dope. There was no cure but more patience.

Two or three weeks later we liked each other again, and though Jack had forgotten his fright I still remembered my ambition to have him come to the lure like a hawk. Most young birds – and animals too – are shy feeders when first taken into captivity. Strange surroundings, strange food and strange methods of feeding combine to make them take the very minimum to keep together body and soul. For the first twenty-four hours they starve. Nothing is effective but force-feeding and I have found that for anything that is reasonably robust it is better to wait and let them come to their food from sheer necessity. Then, after two or three days, they feed even more greedily than they would have done in the wild, because it is not easy to feed them so often.

My jackdaw had long since passed the stage of having inhibitions. He knew my footsteps from all the other boys and would scream in ecstasy even before I came in sight. The moment I entered the door he flew on to my head and shoulders and gaped and yelled and gasped for food.

The shed where he lived was fifteen or twenty feet long and I would crumble a little bread and milk on the floor at one end, fine enough to give him considerable trouble pecking it up. While he tried, I retreated to the far end, showed the yellow yolk of half a hard-boiled egg and he flew to me at once to be fed. A few spilt crumbs would keep him busy whilst I retreated and called him the length of the shed again. This was a game he'd play for hours.

He was very inefficient at pecking and for weeks after he began to feed himself a little he didn't seem able to focus well enough to hit the particular crumb he pecked at. Sometimes he would be half an inch short and at others his beak was still open when it collided with the floor. The trouble, I believe, was simply poor co-ordination between eyesight and muscle for he could see food he liked from afar and had no difficulty whatever in hopping on a perch. At the same time, I am quite certain that it is a common trouble with most birds of his kind. Rooks and jays and magpies that I have reared have all had difficulty in learning to peck accurately. Young chickens and ducks, which are never beak-fed by their parents, on the other hand, can pick up a selected grain within twenty-four hours of hatching.

The very dependence of my jackdaw, however, made him easy to train because he must either come when called and open his beak on command or die of starvation. Within a few days he would fly from wherever he happened to be in his shed the moment I showed him food or called his name. The great moment, therefore, had arrived. I crept from my bed while the rest of the boys were hogged in sleep, collected Jack from his shed and stole out on to the quadrangle with my bird clinging to my shoulder. So far so good. He was frightened by his strange surroundings and my person was the only familiar haven of security in a new strange world. Nothing short of terror would induce him to leave me. I walked majestically round the quadrangle with Jack on my shoulder as proud as a medieval knight with some rare falcon on his wrist. I tried him with food but his apprehension temporarily triumphed over hunger and greed. The first time out he did not eat until he was back in the old shed, which was by now dignified by the misnomer 'mews'.

Morning after morning I was out and about by six. Authority refused to believe that my motive was apparent. My tame bird, it was thought, was merely an excuse. I had never been known to be about early without some sinister objective. Wherever I went I could feel the eyes of masters and prefects boring into my back. I could almost decipher their whispered plots to 'catch me at it'.

Jack, meanwhile, was getting tamer. He looked forward to his hour of liberty and would take off to fly round a few yards and land again on my head like young pilots doing 'circuits and bumps'. Every time he came to my call, he was rewarded by a crumb of hard-boiled egg. Within another week, I felt, he could be trusted at liberty in the certainty that he would fly down to me with the obedience of a falcon to the lure.

Only I had reckoned without those in authority over me. I was just going down on to the school playing-field, where there was space for him to fly and nowhere much for him to land but me, when two figures stepped suddenly from behind the hedge. If they thought me silly enough to smoke so blatantly they were wrong. I was so used to acting with almost subconscious speed though, that they certainly made me jump. And as often happens with wild things, I communicated my fright to my bird. He gave one squawk, soared into the air and flew straight over the brow of the hill.

I was livid and my language left no doubt about my opinion of boys who use the mantle of authority to creep and spy on their fellows. I missed breakfast and chapel and only succeeded in locating poor Jack cowering in the depths of a sycamore when it was too late to do anything about him before class. I spent the whole of break-time locating him in another tree and was lucky enough to tempt him down for some food before beginning my own at lunch. But it was a poor, jumpy, cowering bird that I recovered and it took me a couple of weeks to make up lost ground. By the end of term, however, I could go for quite long walks and my bird would either circle overhead or wait in a tree by my path.

When the holidays came, I took him proudly home and spent hours of my time teaching him to feed himself. Beetles were the best things. He would aim a vicious blow as they scurried across his path and they'd be gone in a gulp. All day long the window of his shed stayed open and he spent much time learning to forage, encouraged by me turning over roots of squitch with a fork. Gradually he became disobedient, taking longer to come when I called him. By the end of the holidays he would eat anything we put on the bird-table but would no longer come to hand. But, in spite of that, he never quite forgot me. By spring, he had a hole in a dead tree and a mate, but even then he would come within a few yards to be fed.

Now, almost half a century later, a colony of jackdaws, which are descended from my original Jack, swarm in every free chimney. Their eyes are as blue as his, the tops of their heads as grey, the sheen on their feathers as metallic. The old chap I reared has long since died, but I shall always remember him with affection because he was the first really wild creature that would live at liberty and still acknowledge me as a friend.

CHAPTER THREE

LOW LIFE

I owe nothing of my love of natural history to starchy school gaffers. Desiccated dons, with knobbly knees and butterfly nets hold no appeal for me.

From childhood, my natural affinity has always been with bad characters, and I learned most of my trade as a naturalist from poachers and gamekeepers and chaps like Hairy Kelly, who never made his mark with social climbers in spite of the fact that he was a true artist, though in an unconventional field.

In modern times, he would be described as a Rodent Operative or Pest Destruction Officer, though he was no unskilled worker, fit for nothing better than filling in forms or stuffing rat holes with cyanide powder or lacing rats' victuals with poison. The mere suggestion would have made him spit with scorn. He wanted no minimum wage for his work because he preferred to be paid by results. When he called to catch your rats, he quoted a firm price of a tanner a rat he caught. Six honest old pence, which would purchase a pint of honest ale drawn from a wooden barrel. A substantial fee for his services, if you convert it to the worthless washers that pass as money now.

Catching the rats was only a beginning. Hairy Kelly was so far ahead of his time that he got paid twice for everything he did. Once for catching them and again when he sold them alive to men with pups to train or who wanted to wager whose dog could kill ten quickest in the pub at night.

The way he accomplished this was well worth watching. He didn't use dogs or sticks or mechanical gadgets devised by scientists. He used his hands. He put a ferret down a hole and, when the rats bolted, he simply grabbed them. It sounds simple and it looked simple because he was so professional.

The secret lay in the speed of his reactions and the length of his arms. He had a reach as long as an orang-outang and the reflexes of an illusionist. However fast the rats bolted, he forecast their movements so that they seemed to dive deliberately into his hands.

Each one he captured was popped down the neck of his shirt and left to cower against his belly till the ferret had finished her subterranean work. The leather strap that kept his trousers up prevented them slipping down too far and the company of the other rats that shared confinement in his clothes filled them all with a false sense of security.

When the excitement had subsided, he transferred his

captives to a sack, which he hung from a branch till he was ready to take them to the local and convert them into currency. He knew that, if you put a sackful of live rats on the floor, they will gnaw through the cloth where it touches the ground. But, if the sack is hung aloft, they will cower in confinement rather than risk falling into unknown space when they cut themselves free.

Hairy was a sallow little bachelor who lived alone, and it was only necessary to get down-wind a bit to gather one reason why, so my family could not understand what attraction he held for me.

His shifty eyes missed nothing. No Red Indian tracker could have unravelled the clues of tiny scratches or faint footprints or bruised herbage to explain the movements of his quarry with more lucidity.

He was the perfect companion to teach a lad his way around the countryside. He knew at a glance if a rabbit hole was occupied, and he could predict which way the rats would bolt for safety when they were chased from their retreat by one of his ferrets.

I, too, gradually acquired the knack of thinking like a rat until I knew, almost instinctively, how to persuade it to seek safety where I could most easily cut off its retreat.

What was work to him, was play to me, so that it was the ideal education for a young naturalist. The one thing we fell out about was Mick, my dog. He was my greatest friend and constant companion. He never left my side all day and he slept on my bed at night. Indeed, if the night was cold enough to set his thin skin into a fit of the shivers, I would find him inside the bed when I awoke.

In those days, Staffords were still matched to fight in the dog pits by the colliers and iron-workers of the Black

Country. They were still famous for the strength of their jaws and for their indomitable courage. So it was not surprising that Mick took to ratting as enthusiastically as I did, and he soon proved more adept at the job than the maestro himself.

There is no jealousy like professional jealousy and Hairy was understandably narked, not least because every rat Mick caught was one less for him to get paid for a second time.

It could easily have been the rock on which our friendship split. Hairy told me, in plain and colourful terms, either to leave the dog at home or to stay away myself.

He had to relent in the end because I had another use. I was the catalyst that made his entry possible to places which would otherwise have been prohibited to him. When a prospective client declined his service on the grounds that his reputation had gone before him, he rubbed his hands together in mock servility and ingratiated himself by saying that he just wanted to show the young gentleman how to go on.

Although, by the age of twelve, the 'young gentleman' knew far more about how to go on than was respectable, the ploy usually worked. So he put up with my dog because the compensations outweighed the snags.

The suspicions of his clients were not unfounded. If we were on a farm, we spent the morning ratting around the buildings, helped or hindered or merely watched by some of the farmer's family. But farmers are sticklers for time. They milk by the clock and muck-out by the clock. And they feed themselves by the clock. Whatever the state of play, when dinner-time came round, they all drifted away to get their feet under the kitchen table.

It was no coincidence that Hairy and I drifted away as well! Not to have our snap but to the nearest rabbit holes, where we spent a profitable half-hour. By the time the family had had their fill, we were back ratting again as if we had been at it all the time. And not all Hairy's bulging sacks were full of rats!

When we had done, and he had been paid his 'tanner a rat', the place was vermin-free. For a short while.

One of the advantages of dealing in live rats was that not all of them were used for training dogs. There were always a few that could be used for a little crafty re-stocking of the premises of clients who paid-up well. Although Hairy could be relied on to keep rats down to an acceptable minimum, it was no coincidence that he was called in again soon.

His knowledge wasn't limited to rats and rabbits – but he was not always right. We were walking quietly down a farm drive one day, when he froze in his tracks and pointed to the gate stump with his eyes. Squatting motionless between the heel of the gate and the hanging post was the most enormous hare I ever saw.

That, in itself, should have warned us. If you spot a hare in a form, and it thinks it is in danger of discovery, it claps tighter and tighter to the ground till it almost sinks from view. Hairy had often told me that if I saw a brick in a ploughed field, and it stayed the same size, it was a brick. If it seemed to grow smaller the nearer I approached, it was a hare!

The hare at the foot of our gatepost didn't shrink. Hairy backed silently away, bent stealthily to the ground and picked up a stone as big as half a brick end. Drawing back his arm with the utmost caution, he took aim and slung his stone dead on target.

44

It would have felled an ox with the force he sent it, and the hare literally disintegrated. Not, alas with the force of the blow but because it had been dead so long that it was in the terminal stage of putrefaction! Nature's practical joke turned the expert to a novice.

To make amends, he taught me his cap-and-stick trick. I'd spotted a hare sitting out in a tussock of grass so he told me to stand still – and watch.

He walked in a straight line, aiming to pass close to the hare without ever seeming to walk directly towards it. When about five yards to one side, he planted his stick in the ground without checking in his stride. He perched his cap on top of his stick.

This riveted the hare's attention and he began to walk round her in a spiral, gradually getting closer at a tangent, but never approaching directly, so that the cap and stick appeared to hold most menace.

I knew that this was a 'good' hare because I had seen her shrink to half her size when she saw us come into the field. As he got closer and closer, Hairy was careful never to catch her eye so that she thought his imminence was accidental. She crouched indecisive, her attention divided between the stationary stick and moving man, who appeared to be about to pass harmlessly by.

It was a fatal mistake. When he was close enough to pick her up, he bent double with the smooth precision of a springing rat trap, scooping her into his fatal embrace before she had realized that her real peril was not with the lifeless stick.

I have copied the trick scores of times with rabbits, which do not even need the hat and stick to distract them, but I have never had the courage of my convictions with a hare.

45

Each time I have tried, a moment's indecision on my part has focused her attention in time for her to duck out of danger before my hands could grasp her.

When I was out ratting one day, I took a pocketful of young rats from a nest as fodder for the ferrets. I later changed my mind and decided to rear one, as I had done with so many other British mammals.

I've always got a few spare cages of one sort or another, so I fitted these five young rats up in a wooden box with a glass front. When they settled I tried to feed them on warm diluted cow's milk, syphoned through a thick piece of wool for them to suck.

They showed no aptitude at all so, as this was a complete failure, we tried a fountain-pen filler. Their mouths were so tiny and, although so young, they panicked so much when we tried to hold them still enough to feed, that this was an obvious failure too. With anything I really wanted to rear, I should have used endless patience and tried at about hourly intervals, for it is very rare to find that young animals will feed within the first few hours of capture. But I was only interested in an impersonal, objective way to discover just how tough the rat tribe really is.

So, before finally giving up, I decided to have just one more try. I put a pigmy electric light bulb in a cigarette tin and, immediately over the bulb, I put a small tin lid of bread and milk. The result was that the warmth from the bulb kept the chill off the whole box and it kept the bread and milk, which was a mere half-inch or so above it, at a nice blood warmth. So that I could see at a glance if the young rats had fed, I filled the lid quite level to the top.

Nothing had happened when I went to bed that night,

but next morning there was a depression in the bread and milk that would have covered a couple of lumps of sugar. I replaced it with a fresh brew and settled down to watch.

First one young rat and then another wrinkled his nose in the direction of the food. They were far too young to sit up and sniff, as their elders would, but they crawled unsteadily over and began sucking the milk from the bread, As they sucked, they pummelled away with their two forefeet to stimulate the flow, exactly as a kitten does when feeding from its mother.

From then on, they never looked back. Within a week they had almost doubled their size and their eyes were open, and I had begun handling them regularly before I went to work and when I came home.

Even rats have individuality. Just as no two dogs or cats behave the same as each other, so it is with wild animals. Of my five rats, one little doe was much more nervous and jumpy than the others. While the others would be prepared to come cautiously forward to sniff my fingers and investigate any new object I showed them to see if it were edible, this young doe would skulk in the far corner of the box. Worse still, she would infect the rest with her fear and, within seconds, they would be as terrified as she was. So she went to feed the ferrets as I had originally intended. Over the next three weeks, three out of the remaining four joined her, two because they were timid and one because she was rather a bad-doer and seemed unlikely to make a decent-sized specimen.

That left the biggest and boldest as the sole survivor. It is usually easier to tame one young animal by itself, than if they are a family. For one thing it will come to rely on human beings for companionship, often developing a definite fixation to such an extent that it loses all interest in its own kind. In addition there is no chance of this devastating hysteria emanating from the nerviest member of the group.

In any case, buck rats are pretty solitary creatures until they grow big and strong enough to fight the adult males of the colony. They are usually found lying quite separate, mixing only to eat and drink and make love. There are, of course, exceptions, as in ricks or rubbish tips or other places that afford exceptionally good cover, though even there, I think, the dormitories are large enough to allow the does to segregate. It is quite essential that they should, because any wandering buck will just as readily eat young rat as he will chicken or pigeon, so that his wives cannot be blamed for banishing him. Old Hairy always used to say that he caught six or seven bucks for every doe, and that that was the only way the rat population was kept in bound at all. If it is true – and I have never been able to prove it for myself, as it would obviously need thousands and thousands of specimens to be sure – then a reversal of the proportions of the sexes to give six times as many females would indeed soon swamp us with sheer weight of numbers.

Be that as it may, the youngster took to us as naturally as a kitten. By this time he was about a third grown and lived in an old bird-cage I'd had for a young thrush. He had long since dispensed with any artificial heat, but I put in a small sleeping-box with an entrance hole rather like a tits' nesting-box. I've always found this a jolly good way of getting any shy animals used to being handled. To catch them in a large cage is very liable to start a minor panic. But if they have a small nesting-box, equipped with a flannel duster or old sock, that they can retreat to, there is never much trouble. They dive out of sight beneath the bedding and scarcely notice the difference when a hand is smoothly slid under them.

My rat never minded being handled by me or my wife and the whole time I had him he never bit me. His cage was on a shelf in a verandah at the back of the house and we discovered that, unless under the stress of sudden terror, he wouldn't jump down anywhere unless he could see how he was going to get back. So we could let him out of his cage at night and he would still be playing about on his shelf when we came down next morning.

Short-tailed field-voles are like this too and, for that reason alone, they are particularly easy to tame and handle. Once they lose their initial fear of man, which they do with incredible rapidity, they can be handled with the confidence that they won't leap blindly off into space and go to ground under the furniture.

As he grew, the first thing that all strangers noticed about him was his shining cleanliness. They began by viewing him with the prejudiced repugnance in which all rats are conventionally held. Then they would suddenly realize that his fur shone with all the brilliant sheen of expensive mink. It ought to have done too, because he was getting bread and milk and cod-liver oil, oats and wheat, toast and meat bones and an occasional raw egg. And there is nothing like raw egg to put bloom on almost any animal's coat, as women who add raw egg to hair shampoo will tell you.

What really intrigued everyone was the way he was always washing himself. He'd sit up on his hind legs like a squirrel, lick his front feet and brush and comb his face and neck and whiskers until you could almost see yourself in reflection.

49

A Weasel in my Meatsafe

He was like Lady Macbeth in the scene where she is sleep-walking, wringing and rubbing her hands and wailing 'Out, out damned spot'. Both of them were simply going through the subconscious motions of washing because of sheer nerves; she conscience-stricken trying to wash off the bloodstains, he scared like a schoolgirl in the presence of strangers. You can try the same experiment for yourself with any of the rat or mouse tribe. Catch one of them alive and put it into a box that it can't jump out of, but which is bare of cover to hide under: as often as not, it will not even stop to run all round in a search for somewhere to escape, before sitting up, in one corner, and going through this almost involuntary wash-and-brush-up routine. Obviously it does not help it to escape; it is not to cleanse itself of foreign smells, for it need never have been handled or touched. My conjecture is that it is nothing more than a severe fit of the fidgets, during the time taken to think up an effective means of retreat.

The next thing most folk noticed about my tame rat was his wonderful 'hands'. They were always clean and pink and dainty, with almond-shaped nails that would not have disgraced a concert pianist. All the women envied them, despite their instinctive shudders. He would hold a nut or piece of toast, or even a tiny grain of wheat, whirling it round and nibbling as it went so that he shed off the waste and ate only what he wanted. It was not until I examined this waste minutely that I was really convinced of the immense amount of damage a colony of rats can do. His powerful central incisor teeth grew so fast that they simply had to chew something, to keep them reasonably short and avoid deformity. He ground them together and chewed bone or wood or bits of his cage, which helped. But he was very intelligent and preferred to keep his teeth in trim on something palatable. He would rather chew wheat grain, nuts or bones. The fact that he could eat but a tenth of what he wasted was of no consequence to him.

For most of my life I have had a variety of animals and birds about the house. (With the exception of childhood pets like tortoises and green lizards and some of the ornamental waterfowl, which live on the pool in front of my house, most of the things I have kept have been British animals or birds, which I have reared or tamed myself.)

One of the facets of this hobby, which never ceases to intrigue me, is the reaction of domestic animals, which I would expect to regard them as intruders. At the time I kept my rat, my old Stafford Bull Terrier, Rebel, was queen of the household. She was the best dog I have ever seen in my life at catching rats. Twice I've seen her catch more than a hundred rats at a sitting; she was as deadly round farm buildings with a light at night as she was threshing or with ferrets by day. Eventually she died of jaundice contracted from a rat bite and I was so upset that I have never had a dog for ratting since, for rats are terrible carriers of jaundice which is very deadly to dogs.

It was natural to expect that she would not take kindly to a tame rat about the place, and I therefore took precautions to minimize the peril to mine. The bird-cage in which he lived was on the verandah shelf by the scullery door. This was so that he would become accustomed to strangers, as the tradespeople constantly passed his cage. So, of course, did the dogs on their way to the garden. Twenty times a day they passed within a foot of this young rat yet neither Rebel nor Muffit, the hunt terrier, so much as wrinkled a nostril.

I tried the experiment of pointing him out to them as he reared up the front of his cage to take a bit of cheese-rind from my fingers. They were far more interested in the cheese. The only explanation I can give is that dogs will chase all 'foreign' cats but share the hearthrug with the house-cat. I suppose that my rat must have smelt just that little bit different from his wild brothers for the dogs to be able to tell which was which. There is, of course, just the possibility that animals give off some specially intoxicating smell when they are afraid and that, in captivity, they rapidly cease to show fear of anything they're used to. Certainly there is evidence enough that dogs can tell when people are afraid of them and are far more likely to attack them than people who find it natural to be friendly. I should have been extremely interested to have let my rat out, where he would have had to exercise his ingenuity to find enough cover to save his skin, and then let the dogs near to see if they showed any more interest then. But I thought too much of him to risk his skin.

Although I kept his cage very clean there was no question of the dogs just not noticing him. I used to cover the floor with three or four thicknesses of newspaper every night and burn it next day. The sort of newspaper we buy is more like blotting paper than vellum, so that every drop of urine was soaked up and, in theory, there should have been no odour at all. In practice, there was a powerful sweetish pong which was clearly obvious, even to my dulled sense of smell, so that it is certain that the dogs must have ignored it. They couldn't possibly have failed to notice it.

Nevertheless I grew very fond of him. He was surprisingly affectionate and would grunt and burble quietly to himself if I poked a hand into his box to scratch and tickle him. If I stopped caressing him, he would hold my finger still in his front paws and lick it until he was satisfied I was as clean as he was. And if I didn't watch, he would pare my nail with his dreadful yellow incisor teeth. In practice, this was perfectly safe, but I always used to wonder just how sharp it would be if he made a mistake. Then my memory would flick back to that childhood day when I was bitten through the thumb, and I'd sweat a bit until he showed his affection some other way.

He had started with me through the accident of forgetting to give him straight to the ferrets; I reared him to see how tough he was and to prove to myself that I was capable of taming something as persecuted and suspicious as a rat; I kept him because we became great friends.

One difficulty I had, though, was with his water. He grew into a fine big buck, gleaming with health and immensely powerful for his size, and his hobby was throwing his eating and drinking dishes about. Since one of these was filled with water every day, the newspaper on his cage floor became sodden unnecessarily quickly. A major problem with all captive animals is to prevent boredom. If they have company of their own species, they are rarely so tame and it becomes necessary to find an acceptable pastime. I tried marbles and bones and scattering bird-seed, which took a long time to pick up. But nothing gave him half the kick as upsetting every dish I put in his cage.

Then, by chance, I was taken over the animal breeding centre of a great Birmingham teaching hospital. Here I saw thousands of white rats and mice being bred scientifically for experiments on diet, and genetics and more mysterious work, which was quite beyond my comprehension.

I wasn't very taken up with the experimental part, but I

was enthralled by the methods used to keep the rats before they were put to any use at all. There was shelf after shelf of wire cages, with metal bottoms, far smaller than anything I would have kept a rat in myself. But they had found by experience that rats were perfectly healthy that way and, as for boredom, neither the rats nor the hospital authorities appeared in the least worried. They had a particularly neat way of making certain that every one of their thousands of rats was handled regularly and thereby kept tame. (They were ordinary white rats and therefore naturally docile, of course.) Every cage was made so that it was impossible to clean it out with the rats still inside. So the attendant had to pick each rat out separately, and pop him into a spare cage, ready cleaned, before he could attend to the old one. The supervisor could then walk round and tell at a glance when any cage was last cleaned and therefore when its rat was last handled.

The tip that I learned, though, was how to supply water. Every cage was equipped with a glass bottle about as big as a sauce bottle. There was a bent glass tube, one end piercing the cork, and the other poking through the bars to the inside of the cage. Any rat which was thirsty, simply sucked the end of this glass tube and a spot of water was allowed to trickle from the bottle directly into the mouth without spilling.

Here was a chance to prevent the soggy mess in my rat's cage, and to test his adaptability at the same time. I took away his water and fixed him up with a water bottle. He took not the slightest notice. So I forced the issue by taking off his diet all bread and milk, raw vegetables or other ingredients which contained much moisture. For a few hours he was nearly frantic. Anyone who has watched wild rats come out on to the top of a corn rick, during a rainstorm, will realize how vital water is to them. They will lick the drops the second they run down the thatch.

My poor rat hadn't got any thatch and it wasn't raining. Furthermore there was a queer glass tube stuck into his cage, which he obviously decided could easily be a trap. He wandered round sniffing everything he knew, but giving the tube a wide berth. He tried to gnaw his way out. He came to me for help. I touched the end of his tube and let him lick my finger, but he wasted my effort by washing my whole hand in gratitude. And then, quite suddenly, he found it. Anything new, that never moved at all, quickly lost its threat for him. He was so busy trying to get out that he forgot all about the possibility of a trap, and brushed his shoulder against the globule hanging on the end of the glass. He licked it off in a subconscious effort to keep spruce, and at once he was galvanized into action. His whiskers were the next part of his anatomy to touch the coveted fluid and then he was on to the trick as quick as a flash. Next day he was sucking away as contentedly as his cousins in the hospital, and I felt a bit smug about it. My rat had the best of both worlds. He had good care and food and a far better life.

When he was a year old, and bigger than any wild rat but the ones 'as big as a cat' you always hear about but never see, we moved his cage to the other side of the verandah. I hung it on the nail where its original occupant, the thrush, used to hang, so that the shelf could be free to grow tomatoes.

It was a stupid, thoughtless thing to do, because I should have remembered that the door was so ill-fitting that he had been able to let himself out at will for months. Whilst he'd been on the shelf, he'd played about till morning and then gone 'home' until the tradesfolk had been.

The first night on the wall he must have come out as usual and climbed on to a broom, standing head upwards, as an alternative to his more spacious shelf. I imagine his weight tipped the balance because, next morning, the broom had fallen on the floor and the rat had gone.

We were able to reconstruct his panic. His one refuge, the cage he regarded as home, hung high out of reach above. He would wander round the strange floor, at first in blind panic, and then with cautious curiosity. It was obvious how careful he had been from the way he'd escaped to the wide world outside. The domestic fowls were free to wander as far as the house, and they had made a nuisance of themselves, coming in through the hole I'd cut in the verandah wall for the cat to use. So I'd hung a sheet of rubber over the hole and the cat had learned to push it aside to come in or go out, but the fowls had not. It didn't fit particularly well and there was plenty of room for a rat to creep past.

My rat didn't though. He'd been reared, from a tiny mite, without any necessity to be frightened and suspicious. Yet the very first time he had been cast on his own resources, he had been so scared to take a risk with this queer flap, which might be a trap, that he had gnawed a fresh hole in the woodwork about six inches away instead. So all we found at breakfast-time was a fresh rat hole and no rat. We searched high and low without avail and put his cage in the garden in the vain hope of finding him back home next morning. That night we wakened to hear a rat squealing in the garden. Too late I pelted down to the rescue. A neighbour's cat was just disappearing over the wall.

CHAPTER FOUR

BILL BROCK

At school I remember two books in particular, neither of which appeared in the curriculum. One was a collection of short stories by Henry Williamson called *The Old Stag and Others*, and the other was *Diana, My Badger* by Frances Pitt. In Williamson's book I was particularly taken by a character called, if I remember, Bloody Bill Brock, the toughest wild badger which was ever tormented by a terrier. Miss Pitt's main character was a tame badger she had reared herself.

Since those days, as I have gradually got to know more and more naturalists, I have been struck by the almost universal attraction which badgers have. I know people who spend every available moment perched in a tree near their favourite badger sett making scientific notes or taking photographs; I know men and women who have kept badgers in captivity for varying times and with varying success; I get frequent letters from folk who would like either to watch or to keep them and enlist my help.

My boyish imagination was fired with an insatiable appetite both to know and keep a badger. I watched them in the wild, seeing fleeting glimpses of their characteristic black and white striped heads and greyish rather ill-proportioned forms, but it was years before I achieved a lifelong ambition to own one.

I had noticed a photograph in my evening paper of a tough-looking gentleman holding up two dead badgers like a big-game hunter of the last century. He wore a self-satisfied smirk that put me off him from the start. A corner of my mind registered the fact that his name was Gripton. A hunting farmer who liked a bit of sport, either orthodox or otherwise, later told me where he lived and that, if I wanted a young badger, he was my man. The temptation was irresistible. The next Sunday afternoon found Bert Gripton and me warily sizing each other up.

I'd met lots of chaps like him, I thought. He was about my own age, powerfully built and looking the part in a tattered old polo sweater and breeches and wellingtons. He was surrounded by a motley collection of terriers and whippets and at least one litter whose parents had obviously contracted a mixed marriage. Right from the start I could see that he could tell a good tale. 'This was the best terrier in the county at drawing a fox.' 'That whippet could catch a hare.' 'This one had won its weight in cups in the show ring.'

He was equally suspicious of me. Every Sunday afternoon he was bombarded with questions and theories by chaps from the towns who tried to impress him with their knowledge but had never been much nearer to a badger or a fox than a picture in a book.

By the time we finished up, I think, neither knew quite what to make of the other. My old bull terrier had just died, so we'd done a deal and I'd bought one of his whippet lurchers. She is a delightful little bitch who is lying full length on a settee with her head on my knee as I write this. I have certainly never had a more intelligent or better bitch in my life. As we parted, he promised to let me have the first badger cub he caught next spring. It is, of course, illegal now either to dig out badgers or to keep them as pets, because of the 1973 Badgers Act, but the next February, there was a phone message to say that a badger was waiting for me to collect at Bert Gripton's, and would I go that night for certain?

It caught me on the hop. I rushed round to the chemist's for a baby's bottle and tin of baby's food. I left the question of suitable pen till I'd seen the size of the badger and within half an hour I was twenty miles away at the Griptons'!

Bert's kitchen was tiny and well filled by himself and his wife, his daughter Ann, his father-in-law and his mother-in-law. There was usually a terrier or two on the hearth and if there weren't any cats it was probably because young Ann had tipped them out so that she could play with her tame mice on the table. When I walked in, the walls bulged and the shadows from the lamp shifted over a bit to make room for me.

It was a simply foul night and Bert enquired if there was any fog and if it had been inconvenient coming over. I asked after the family and what sport he'd had lately. The one thing I really wanted to know was where was my badger and how was he. Or was it a 'she'?

I was simply itching to know and yet, for some subconscious reason, the words just wouldn't come. We were still fencing with each other and I knew that everyone in the room was watching me to see if Bert had called my bluff or if I really did want a badger as much as I said I did. They were speculating about my chances and I wondered what they assessed the odds to be before I committed myself in words. We played this ridiculous game of bluff, it seemed, for an eternity. Then Bert took pity on me.

On the centre of the table, right by the paraffin lamp, there was an old cardboard grocery-box about as big as a boot-box. I'd seen it and the thought flicked through my mind that, if that held Ann's mice, they must be pretty warm, what with the fire roaring half up the chimney and the heat from the lamp and the fact that there were half a dozen steamy bodies in this tiny room.

It didn't hold mice though. It held my badger. With all the artistry and melodrama of a conjurer doing the star trick of his performance, Bert plunged in a hand and plucked out a squirming bundle of helplessness, no bigger than a tiny kitten.

I took it gently, almost reverently, from him and cupped it in my palms to hold it still whilst I examined it. I needn't have bothered. From above it looked a perfect miniature badger. Grey back and sides, and head with the characteristic black and white longitudinal stripes. He felt oddly smooth and warm. My first reaction was that he'd wet on my hands from fear but then I noticed that he was still blind. His two queer piggy little eyes had never seen the light of day. I turned him on his back to confirm what my sense of touch had already told me. Sure enough, 'he' was a little boar, but what did surprise me was that he was completely naked and bare underneath. What I had taken to be wetness was merely the warmth of his naked tummy.

That explained the almost tangible sense of doubt. Nobody in that room thought it possible to rear my cub. It was bad luck on that badger, of course, but at least it would put me in my place and show what I could do – if anything.

I accepted the unspoken challenge. I'd reared plenty of things before and I was determined to rear this one. More important than food was warmth. So when I got to the car I took the cub out of his bag of hay and stuffed him down my shirt against my warm tummy. He was just the size of a mole and, young as he was, his front feet were already powerful. I thought he was going to burrow through me before he got comfortable and many a time since I've wondered what the authorities would have thought if I'd had an accident going home and landed in hospital. It would almost have been worth it to watch the nurses' faces when they'd undone my shirt.

We got home safely though and my wife was as delighted as I. Only when we examined our new treasure by revealing electric light we found a long red scar where he had obviously only just escaped from the terriers with his life. It wasn't a very good start. I remembered the knowing looks at Gripton's and decided that the first thing to do was to christen him. No infant should be allowed to die unnamed.

So he became Bloody Bill Brock after Williamson's far-off character.

Next we tried to feed him. It hadn't occurred to either of us that he would be so small, but now we saw the size of his diminutive mouth, a baby's teat was quite out of the question. How I wished for a cat with kittens which might have been persuaded to take on the dual role of giver of food and warmth.

We tried, singularly unsuccessfully, with a fountain-pen filler, but quickly gave up for the night. It is nearly always impossible to get young wild things to feed the first few hours they are in captivity. Provided they are reasonably strong, I never bother, but leave them to their sulks in the sure knowledge that hunger will bring them to their food within twenty-four hours.

What is essential, though, is warmth. At one time I used to mess about filling hot-water bottles and replacing them at regular intervals as they cooled off. I always found it quite hopeless. They were much too hot to start with and when they needed replenishing in the small hours I was usually too solidly asleep to do anything about it. And when I did eventually filter back into consciousness, my charge would be shivering and whimpering with cold.

By the time I acquired Bill Brock I had developed a much more efficient technique. I took the largest, flattest cigarette tin I could find – I think it held 200 – and covered it with an ordinary sock. Inside the tin I put a 15-watt pigmy electric light bulb. The result was ideal. The bulb was just strong enough to produce a heat in the tin which was the perfect imitation of a mother's furry tummy. It had the incalculable advantage of being as warm in the morning as it was the night before. Placed in the bottom of a box it gives off the exact amount of heat which seems to be required by most young things, and I have used it for badgers and squirrels and rats and even a Siamese kitten.

In passing, I would mention that it once got me into severe hot water. I had been demonstrating on television how easy it is to rear chickens on my cigarette tin and pigmy bulb method. Knowing from bitter experience how many people have nothing better to do than look in by the hour, for the express purpose of criticizing what they see, I took normal evasive precautions. I said that any children who thought of trying it should get their bulb and box fixed up by some knowledgeable grown-up. That, I thought, should put me in the clear. Not a bit of it. An irate electrician (probably a shop steward!) wrote to his local paper to say what a disgrace he thought it was that the BBC should employ amateur madmen to teach young innocents to electrocute themselves. And I spent the next few hours – which happened to be from about 10 p.m. till 2 in the morning – telling all the other newspapers what I'd really said.

Bill Brock, however, was not so critical. If he didn't like our baby food or our fountain-pen filler, at least he appreciated warmth and soft flannel. My wife had produced an old blanket with which we had completely lined a small packing-case. The cigarette tin with its little bulb was stuck under this blanket at one end of the box. Young Bill Brock, therefore, had blanket to lie on wherever he scrambled. So long as he stayed at one end on his tin, he had a warm blanket, as warm as if he was curled up with his litter-mates.

I think that first night was as restless for me as it was for him. Ever since I'd read those first natural history books at school my ambition had been to own a tame badger. Now I'd got him it seemed he was rather too young to do any good. If he died, I should have felt partly responsible for his being taken from his mother, for the receiver is deemed more guilty than the taker. I should also have felt extremely sheepish the next time I saw Bert Gripton who, I was certain, thought I wouldn't succeed.

Next morning it was cold and grey after a bitter night. But for once a warm bed didn't pull. I crept into the kitchen and slid a furtive hand quietly into the blanket in my packing-case. Joyously I discovered that the piggy little form on the tin was warm and wriggled and snuffled. So far, so good.

But he wouldn't take any baby food. We tried poking a fountain-pen filler in his mouth but he only blew bubbles. We tried the baby's teat. As a last forlorn hope we dipped his nose in it. The only result was that we got him rather sticky and had to sponge him clean. Normally, I wouldn't have worried at all. So many wild things are wild and obstinate at first but act as though they have always been domesticated as soon as they feel the pangs of hunger. But my young Brock was different. The stakes were so much higher.

That night, when I got back from work, my wife was elated. She'd mixed some baby food at midday and he'd taken about a teaspoonful from the fountain-pen filler. I was all agog to try for myself. Taking the rug and badger, I settled down in an easy chair to fondle him until his milk food was warm. It was indeed a prophetic pose. For the next three or four weeks I was to spend at least two hours each day in this same chair whilst we really got to know each other. This first time, however, I wasn't very successful. Despite his hunger he hadn't taken to the flavour of his new food and his mother's breast must have been far more shapely and attractive than my glass tube. After about twenty minutes' struggle I succeeded in injecting a tea-spoonful or so of nourishment into him and felt that he must have lost more strength in the struggle than he had gained in the meal. At night, things were but little better and we all went rather disheartened to bed.

When I got down next morning, he started to chatter and scream as soon as I touched him. He nuzzled my hand and was obviously nearly famished. The very smell of warm baby food sent him into such ecstasy that I was afraid he'd break the glass tube with sheer enthusiasm. His delight was short-lived. He sucked and spluttered his food down until he was partly filled and partly exhausted. Then his joy subsided into sleep, whilst I began my own breakfast with better appetite than I'd had since he came. My one worry was the danger of putting a rather frail glass tube into the mouth of such a tough young animal.

He was very difficult to start feeding from the bottle and it took a lot of experiment before he would accept a normal teat from a human baby's bottle. If the hole was too large, he would gulp the milk down so fast that he nearly choked and, if it was too small, he soon gave up trying.

One of the most critical things of all was the fact that it didn't take long for a teat to get shredded by such a vigorous youngster, but he would almost starve before he would accept a new teat with an unfamiliar smell. So I had to take great pains to try to get him using at least two teats and I usually succeeded in 'breaking the next one in' by offering it to him first feed in the morning when his appetite was sharpest.

Even so, when he had taken the edge off his hunger, he

would often stop feeding and sulk, so that I had to swap teats half-way through the feed and finish on one of his old favourites.

Each feed took about forty-five minutes so that we got to know each other pretty well before he was weaned, which undoubtedly helped to ensure that he stayed tame for the rest of his life. But forty-five minutes before work in the morning is a fair slice out of the day.

To badgers who have nothing else to do, I suppose, this wouldn't have mattered. I had to be shaved, breakfasted and nine miles away at work by nine in the morning, so that it affected me quite a lot. I fed him before breakfast, about six in the evening and just before I went to bed, and my wife fed him at mid-day. And, although we used to grumble we were so elated by our success that our complaints were nothing but a matter of form.

Within four days, one piggy brown little eye opened, but not the other. He retained this evil-looking leering wink for the next couple of days and then suddenly he was transformed from fat immaturity into a perfect replica of the pictorial badger which ambles through the pages of books the world over.

Most young things grow very rapidly before they are weaned, and he was no exception. His belly became covered with black fur, he became clumsily playful as puppies do and his capacity for baby food increased even faster than his stature. As his mouth grew, it became possible to substitute the glass tube and doll's teat for a normal baby's bottle which was much easier to keep sterilized. But as he grew larger and needed more food at each meal, he seemed quite incapable of taking it faster, so that every meal still lasted about three-quarters of an hour. Despite this, he was

66

most amusing to feed. If the hole was too large, so that the milk flowed freely, he choked and simply stopped feeding. If it was too small, he became simply furious and dived his head between his two front legs exactly as an adult badger does to dodge punishment. The result was dynamic. The teat stretched like a catapult, until the tortured rubber either pulled off the mouth of the bottle or slithered from the young animal's almost toothless jaws. In either case the result was the same. I got a lapful of sticky baby food.

From the very start, I used to let the dogs come whilst feeding was in progress so that they would get used to the sight and smell of badger in the house. My friends who came for an evening used to cock a knowing eyebrow when I slipped out of the room at about ten o'clock and returned laden with baby's bottle and towels and rug for protection. So many folk are so constipated with convention that they regard the idea of having a badger in the house as more than a little eccentric. They were patronizing in their interest until inevitably, they fell under the spell as soon as they really made Bill's acquaintance.

Then, quite suddenly, they would ask to be allowed first to stroke him, then to hold him and eventually to feed him. It would take some time for them to acquire the art of slipping his teat into a mouth rapidly disappearing between his forelegs. And each failure meant a spatter of gooey milk on waistcoats and dresses unsullied by anything less civilized than gin or sherry. I would snuggle back in my chair and watch with almost paternal pride, waiting for the cub to finish feeding and shuffle backwards a few steps. This was the signal that my guests were about to be sullied with something even more elemental than baby food, and I took a detached and objective interest in comparing

A Weasel in my Meatsafe

their reactions. Most would giggle and dab self-consciously and return to their more conventional hobbies; occasionally, however, someone would become infected by my disease. Its symptoms are easily recognized. An interest in any wild thing is so absorbing that neither time, nor discomfort, nor convention matter by comparison.

This question of time is perhaps more important than any other single factor concerned with the taming and care of wild animals. A very great friend of mine, who is the best veterinary surgeon I know, once summed it up for me. 'The major requirement of a good vet,' he said, 'is to forget all about the clock.'

The critical time was over, but I have never found anything so difficult to wean as young Bill Brock. Normally, when most young animals I've been caring for obviously grow strong enough to fend for themselves, I give two bottle feeds a day and offer two more sloppy ones. Hunger is the finest sauce there is and, after the first day, there isn't usually much bother. I was so very proud of Bill, though, that I wouldn't take any chances. Besides, I knew that the longer he was bottle-fed the tamer he was likely to be. The result was that in May, when we wanted to go for a few days' holiday, he still wasn't properly weaned. All through March and April he'd spent over two hours a day on my lap or my wife's being fed from his bottle. It is true that by this time he was quicker drinking his food, but he was also more playful and agile. So what time we made up on the actual feed we lost again because we enjoyed playing as much as he did. He wrestled and bit exactly as a puppy does before he cuts his second teeth. Only he was rougher and 'harder mouthed' than any dog I've tried, except a Stafford Bull Terrier.

Our holiday was a mere week away, so something had to be done. If we missed a feed and offered bread and milk or bread and baby food, he squealed and chattered with temper and stayed hungry. Then we discovered that half a round of bread all but floating in a tin of warm milky food worked miracles. He smelt the coveted liquid and nuzzled the spongy bread as if it were the sow badger's belly and he was searching for a teat. As he nuzzled he sucked, so that within a day or so he was weaned and would feed perfectly satisfactorily from a bowl. Now he no longer held any terrors for the friend who had promised to care for him while we were away. A week later, when we came back, he seemed almost grown-up. He squealed and chattered with delight to see me again, and I gave him a bottle as a reunion feast.

Quite suddenly the problem arose of where to keep him. I was very lucky because I lived in an ugly red brick box of a house so beloved by the Victorians. As a badge of their eminent respectability, they had provided it with an excellent range of outbuildings, including a stable. I'm never very fond of cold brick floors to rear young animals, so I covered this stable with a really thick layer of finely-chopped straw, and I kept the same packing-case sleeping-box, with its false bottom, to accommodate the little bulb for heating it. In order that he could gradually be weaned of artificial heat as well as his bottle, I supplied a spare sleeping-box with more straw and comfort, switching off the current in his old one during the daytime. Only as the weather warmed and he took to the more softly padded (if cooler) box, did I cease heating the original one.

The hard work was over now and the fun began. Our young badger had spent so much time with us that he was completely unafraid. He came out in the garden after tea and gambolled round, either with us or with the dogs. When he got excited, he chattered and squealed and his long coat stood on end like a golliwog so that, quite suddenly, he looked twice the size and very formidable.

But he wasn't a very good gardener. While he was the size of a small cat it hadn't been so bad but by the end of June he was bigger than Muffit, my hunt terrier, and literally ploughed through the flower borders, leaving them with a morning-after-the-tornado look. I couldn't see that there was much more left to spoil, and 'Anyway,' I told my wife, 'everyone down the road has got a barrow-load of flowers in the garden but we're the only ones with a badger'. It didn't work. Bill and I were asked to go and play in someone else's yard.

Across the road was a field belonging to my father, so we went there. On the way we were constantly stopped. I was generally regarded as slightly 'odd' by the neighbours, so they had no qualms about accosting me to see 'what I'd got this time'. They mostly thought Bill was 'pretty' or 'quaint' and made the same sort of clucking remarks about him that they would have made about the latest baby in the road or, at any rate, about anything as unusual as twins. And one good lady was quite certain he was dangerous and wondered if her children were safe.

Once the field gate was shut behind us, all was quiet and peaceful. I discovered that, wherever I walked, Bill Brock followed as close to my heels as a trained spaniel, for the very simple reason that he was simply terrified of being lost. He was so short-sighted that he couldn't see me five yards away unless I moved. Often, when the dogs were playing with him, they'd all run 50 or 100 yards across the field, and then the dogs would put on a sprint and leave him standing. His look of horror was eloquent.

With his legs four-square and rigid he'd stand stock-still and petrified, straining every muscle to catch the faintest sound to indicate which way to run to me for protection. Sometimes I'd tease him and freeze as still and silent as he was. Seconds would grow into minutes and neither of us would give way. Then he'd begin to move. It was a most unexpected gait and almost impossible to detect the precise instant when he slid from his statuesque immobility. A few minutes before, when he'd been playing with the dogs, there had been a puffing and pounding of feet enough to sound like a herd of cattle stampeding. His instinctive caution had been conquered by his tameness.

The moment he was lost, he reverted to a wild animal which could steal over the ground as smoothly and as silently as a shadow. His whole natural intuition took charge, for his situation must indeed have been terrifying.

With sight as short as his, the flatness of the field must have seemed as immense as an ocean; the problem of which direction to take to find cover or myself, the one familiar trusted object in his strange new world, must have seemed insoluble. His instinct was quite wonderful, however. He would cock his nose in the air and test the breeze and then he would drift off across the prevailing currents like a sailing-boat tacking across the wind, or a pointer 'working' a moor for grouse. The whole time his sensitive nose would be probing and testing every eddy of wind for the faintest familiar whiff. If he drew blank, he would turn on to the opposite tack and drift round in a semi-circle until, sooner or later, he would arrive down-wind from me. His delight then was infectious. Caution would be cast to perdition and he would come galloping boisterously across the turf, until he fetched up, panting and whimpering with delight, at my feet.

His curiosity was quite insatiable and he often wandered just as far off without the stimulus of a game with the dogs. This time, though, there was never any difficulty about the return. Every few yards he would squat on his haunches, as if for a few seconds' meditation. He appeared to be trying to memorize his route and his surroundings.

He was doing nothing of the sort. Badgers belong to a group called Mustelidae, the common feature of which is a musk gland under the tail at the anus. It is this gland that allows the skunk to give off such an insufferable stench and has earned the polecat the reputation for stinking. Stoats, weasels, otters and martens all belong to the same group and are liable to smell excessively under the stimulus of fear or excitement.

Badgers use this gland for normal, less obvious purposes. When Bill was on strange ground, he squatted every few yards and pressed his anus to the ground, rather like a puppy with worms. Each time he made contact, a tiny drop of his musk was left, so that when he wanted to return, he merely had to follow the trail he had laid as easily as if it were a paper chase. When he was a cub, it was quite fascinating to watch him gradually learning his way about the field by this process of retracing his own footsteps.

That wasn't the only use for this gland, either. It was stimulated by fear or excitement, and boisterous games with the dogs often left behind a strong, rather sickly pong, which hung almost as tangible as mist on a September morning. It was used, too, as a symbol of possession. Whenever I went into the stable to see him, he quite invariably squatted on my foot. For a long time, however carefully I examined my shoe, I could find no trace of any liquid deposited on it. But then I noticed sometimes a tiny drop about as large as a pinhead. This minute spot of scent was enough to warn all other badgers that, wherever it had been deposited, was private property and that trespassers were well advised to keep away.

His play with the dogs included a great many mock battles. He had a trick of diving his snout between his front legs so that he presented the top of his neck and shoulder for the dogs to bite. As he grew, this became an almost invulnerable mass of muscle and sinew so that, even if their worrying had been in earnest instead of play, he would have suffered no damage. The dogs, however, were lucky he was not serious. As soon as their attention was riveted on the expanse of neck muscle he offered, his snout shot from behind his front leg and he had them by the forelegs. Just one rapid, crippling snap after another simply showered out at the dogs' legs, feet, snout, ears and neck from this least expected quarter.

The more furious the game waxed, the better he enjoyed it, though he never got rough enough to cause any real damage. For minutes on end, the contestants rolled and snarled and puffed and worried. Then, as if at a signal the dogs would disengage and tear off, with behinds tucked in, through circles and figures of eight all over the field. Bill Brock was no less spectacular. When he was in no hurry, he ambled in a smoothly deceptive bear-like gait, which soon covered a lot of ground. From the side it looked as athletic as it was, though from behind his back legs looked weak and curiously like a hedgehog's. When he was playing, however, his movements were dynamic. His back arched and his legs became stiff, so that he propelled himself in an undulating motion, like a baby rocking-horse.

Such strenuous exercise soon produced exhaustion. The dogs would lie and laugh with lolling tongues, for then it was my turn for punishment. Bill had discovered how vulnerable my ankles were and would steam in to the attack, so that he had me dancing evasively from one foot to the other, until I had to gather him up in my arms for the sake of peace and quiet. That was his attack and defence routine. Just as lads play soldiers in a subconscious effort to equip themselves for any adult eventuality, so most young animals indulge in mock attack and defence, until the least sign of danger places them quite automatically in the very best posture for self-preservation.

The next most important need was food. By this time, his diet had been augmented to include quite a variety, ranging from raw egg and raw meat to honey and bread and treacle. In the field he learned to fend for himself. He used his nose to root like a pig, so that it became quite obvious why the neck and shoulder muscles of badgers are so much stronger than the hind-quarters. But his nose was even more sensitive, and he could detect bulbs and other delicacies several inches below the surface. Then he would give a couple of strokes from his immensely powerful front paws and expose the object of his curiosity. But that was often all the interest he'd show. The very sensation of digging was so attractive that, once started, he would forget his initial objective and set about tearing up great mounds of earth from the sheer love of showing his strength and practising his muscles.

I found it intriguing to spend a few evenings watching

wild badgers at the same time that mine was growing up. There were two very strong setts at the hotel I stay at in Wales and quite a good sett within four miles of home. In each of these there were young badgers of about the same age as mine and I wanted to compare their reactions.

Both setts were placed on the side of a hill amongst fairly thick rhododendrons, though the one near home was not on nearly such a steep bank as the Welsh one. I used to go in the afternoon, if I could, to test the direction of the wind, so that I could approach from the safest direction in the evening. In most of the books I have read about wild badgers, it sounds easy to hide up a tree or on a vantage point, so that the wind always blows from the emerging badger towards the teller of the tale. I must have been unlucky. Go to the nearest badger sett you know on the stillest of days and exhale a lungful of smoke. I will gamble that it will eddy and swirl, so that from anywhere within twenty or thirty yards wisps of it will inevitably float to any badger emerging from a hole. And a windy night only increases the danger-zone. Badgers always seem to choose sites with this desirable feature.

In June and July, however, I found the young wild cubs almost as guileless as mine. I used to sit thirty or forty yards away until I heard them come out and start gambolling about. Then, during each fresh bout of play, I would creep a few paces forward. The air would be festooned with the heady musk of their excitement and I have often approached almost near enough to touch them.

Their play with each other was almost exactly similar to my young Bill's. Only he was fed so well that he played longer before breaking off to feed himself. Like him, the wild ones would dig little, open latrine-pits, which they would use like cats but never trouble to fill in. Like him, they would forage by the hour and were even noisier gourmets. Sometimes I would wander round the hedges and fields next day to try to establish what the wild ones were feeding on, so that I could offer Bill the same fare. And as the summer wore on, I often came across evidence of them opening up bee and wasp nests, though I never saw them at it.

Finding a bumblebee nest in the hedge, I decided to try for myself. I had got Bill used to coming out on a dog collar and lead and I attached a six-yard ferret line to the lead, so that I could control him in the event of any really effective resistance.

He passed the hole which led down to the nest without taking the slightest notice. Then a bee flew in. Immediately he froze in mid-stride, with forefoot raised in question. His sensitive ears were obviously attuned to the subterranean murmur. In an instant, his veneer of civilization vanished again, and a grey shadow drifted back up the hedge where a young badger had passed a moment ago. The hunt was on and his delicate sense of smell had joined forces with matchless hearing. Without a second's hesitation, his long, sensitive snout probed down the hole and his ribs bellowed gust after gust of intoxicating odours.

I never discovered quite what happened next. Certainly Bill began to grunt and snort and his whole coat frothed up like a mad cat's tail, and he started to dig as if all the devils of hell were behind him. But I can't make up my mind whether his driving force was the intoxicating smell of newly discovered honey, or if he was literally stung to action.

Within two or three minutes, half a dozen bees were

harmlessly entangled in his puffed-up fur and the hole was big enough to cover his head. Then, as suddenly as he'd begun, he stopped. He shook his head and pawed his snout. All was not sweet that buzzed, it seemed. There were lots of things he could think about that were much more fun than digging out silly bees' nests. And nothing I could do would persuade him otherwise. Whether badgers do not become tough enough to ignore stings until they are adult, or whether they only take bee and wasp nests when they are less well-fed than mine, I have never discovered.

I now believe that badgers dig out bee and wasp nests mainly during the hours of darkness, when the insects are quiescent and less likely to identify and attack the culprit that has desecrated their nest. That is only surmise but the one thing I do know for certain is that Bill preferred his honey from a jar!

There was one other major difference between him and his wild relations, and that was his reaction to noise. Any unusual noise at all will send wild badgers diving for the safety of their holes. I could hammer, saw, chop wood, or, as an experiment, even bang on a tin tray, without his taking the very slightest notice. But if I shuffled my feet slightly on gravel, or broke a match-stick, he would dive for cover as if he had never been tamed. It was as if he regarded me as a clumsy oaf who could be expected to make the most heathenish clatter. But a tiny, unfamiliar sound, not associated with me, would penetrate his subconscious memory and produce a purely automatic reflex escape routine.

His capacity for adapting himself to unfamiliar conditions never ceased to astonish me. I was particularly anxious that he should grow accustomed to strangers, so that I used to take him four or five miles by car to a favourite pub at Wednesfield. He quickly became almost as fond of the car as the dogs are, and learned to lie in the back, with his head between the front seats, so that a mere twitch of his neck brought his muzzle into reassuring contact with my familiar hand or leg. When we got there, I used to pop him under my arm and slip in the back way to avoid the gawping eyes of loafers in the street.

Inside the bar, the 'regulars' were well used to my eccentricities and took rather a pride in greeting 'their' badger. Mr Gregory, the owner, presided in friendship at one with his guests, without the formality of a bar to divide him from his customers. Two walls were lined with shining bottles and glasses, with the beer pulls by the door. The other two walls had seats along them, being broken only by the fireplace. The dogs and Bill used to play on the cloth rug in front of the huge fire, while the rest of us watched in admiration till the novelty wore off, and then retired to our beer and our gossip. When he was tired, Bill would sometimes lie down with them but more usually wander round until I picked him up and allowed him to fall asleep on my lap. So long as he was going there pretty regularly, he didn't take much notice of the people in the room, unless one shuffled his feet on the tiles, or made one of the sudden small noises he hated so much more than the really loud ones.

What we used to pray for was the arrival of a stranger. The routine was always the same. He'd call for a drink and retire to his bench to size up the surroundings. We'd stop talking and glance slyly at him, till he got a bit fidgety and wondered if he'd got a button undone or something. Then we'd all start talking at once, until the whole smoky air filled with the rather self-conscious sort of chatter that always rolls out of an awkward silence. We were only play-acting, of course. Repeating a routine that had once been accidentally successful and never failed afterwards.

The result was always the same. As soon as our victim thought we'd forgotten him and were submerged by our own chatter, he'd allow his eyes to wander haphazardly about the room to assess what sort of community he'd landed amongst. And he must have found us an ordinary, uninspiring bunch. Then, as his gaze passed me to some rather interesting bottles by my shoulder, his eyes would stutter and flick back to my side. The odd black and white head on my lap just didn't seem true. That was the interesting bit, for we found we could predict the reactions of humans drifting in, as accurately as we knew what wild animals would do under given circumstances.

Every single time a stranger saw my badger on my lap he glanced away self-consciously, as if we'd caught him looking at an imaginary pink mouse. The conversation disintegrated into drivel as we concentrated on our victim. He never noticed though, because he was always too busy mentally pinching himself before he could summon up courage to look again. He had to be quite certain that no one was watching him first, just in case he was making a fool of himself, and then, as if by coincidence, his eyes would drift across my knees again. Sometimes it would seem an eternity before he was convinced that he could really believe what he saw. Then we'd buy him a drink as recompense for pulling his leg and, as often as not, he'd bring his own pals in later on, to have the same joke played on them.

It was all very good for Bill Brock as well as for trade.

He got used to gales of laughter – which most animals hate – and strange voices and, above all, to sudden movements. Through years of handling various sorts of animals, I have learned to control jerky movements subconsciously. I often notice that animals I think I have tamed are as wild as ever after two minutes in strangers' company, not so much because of strange smells and voices, as because staccato movements cause them to panic.

It was through taking him to the pub, though, that he inflicted his one and only bite on me. The Press heard about this badger which went to the local and sent a photographer over to get a picture. He was quite the slowest creature that ever worked a camera and expected me to keep my ten-month-old badger still for about ten minutes while he focused and fiddled with light meters and went through all the abracadabra of his kind. Bill's manners were slightly more primitive than mine and his impatience became so obvious that I couldn't hold him still at all. So I went for a piece of fruit cake which he adored. Waggling it in front of him to make him look in the right direction, I was not quite quick enough. He struck at it like a rattle-snake and got the fleshy bit between my thumb and fore-finger as well. The cake was exactly what he wanted and he clung on like a bulldog, so that it was a good minute or so before he let go. The only thing which stopped me express-ing my opinion of all fools with cameras was that I believed it impossible to make the chap realize that the bite I'd just sustained was his fault and not my badger's.

I say it was the only bite I got. It was the only serious one, though between the age of six and twelve months he made me bleed, on average, three or four times a week. I'd got him properly weaned and he was eating a very varied diet. He had a duck egg every night for foundation and made up with raw rabbit, or any fowls that died, or cooked potato, or bread and milk, or bread and treacle. He'd grown very strong and got so rough in his play that it was only just possible to compete without using heavy gloves. I believe that lots of people rear badgers to about six months old and then give up. They have an 'accident', or 'escape', or 'die', or 'have to go to the zoo'. Anything except stay in captivity.

The reason, I think, is twofold. During the first few months, the owner takes enormous pains, until the novelty wears off and he thinks he can supply food every day but need only spare time when he feels like it. The second reason is that, as young badgers grow up, they pass through a very rough phase which is easily mistaken for savagery. I dealt with this by never missing a single night playing with him.

People who claim to have kept badgers loose indoors never cease to astonish me. We keep as free a house for animals as most and Bill often came inside with the family. But only under supervision. If he heard a spring creak he'd think nothing of ripping the chair side clean out with his powerful claws. They're so powerful that I even had to line the loose-box door, where he now lived, with steel sheet, as he could go through a wooden door in a night. And there was once a loose quarry in his floor. One night he managed to work it out and by next morning he'd fetched a dozen more up and got a hole under the floor like a rabbit warren.

The first winter we had him was rather a worry. From September onwards he'd put on an enormous amount of fat. I did quite a lot of experiments at that time to find exactly what his food preferences were. Every time he'd eat the sweet things first and raw meat last, if at all, and he was very fond of egg, partly because he loved to hear the shell crunch, and almost as fond of bread and milk as cake.

While I was at it, I put the famous drinking question to the test. Lots of people say it's impossible to tell if a badger sucks or laps, and for a long time I couldn't make up my mind myself. Now I'm certain that he did both. There is no doubt at all that he lapped, that was as obvious as with a dog. He was an exceptionally noisy lapper and made the same sort of sucking noises as a pig as well.

By Christmas he went off his food. I wasn't particularly worried, because, although I knew wild badgers didn't hibernate, as I often tracked them in the snow, I did expect a definite slowing-down of his general rate of living. Then he started to cough.

Now I'd heard that captive badgers commonly died from a rather obscure throat infection and I naturally became rather worried. It became impossible to tell whether he was taking as much food as he needed to keep out the cold, or whether his cough was putting him off. Since he would eat nothing I couldn't give him medicine – you can't very well 'drench' a badger – and anyhow, neither my vet nor I knew what to give him. So I put a couple of spots of cod-liver oil on his snout every night, left him to lick it off and hoped for the best.

Cough or no cough, he came through the winter in fine form and never had any recurrence. He went from about mid-December to the end of January or middle of February eating only on average once a week. He was about forty-five or-six pounds before he started this fast and about thirty-seven or-eight when he came back on to his food.

I have several times tracked wild badgers, from the time they have left their setts in snow until they have returned, without finding any traces of a meal. I don't think this means that they are having a hard time, but simply that it is natural at this time of year to do without food. Since they are putting practically nothing in, it is not surprising that their latrines are virtually unused and the one thing that I am particularly anxious to discover is why they leave their setts at all, instead of conserving their energy like animals which undergo a conventional hibernation.

That first autumn, while he was still fit and fat and active, I was asked to show him to the children on television. I have said that, to put it mildly, he was a bit rough at the time. Looking back I can see that that is an understatement. I could play with him for a certain time, then he became so excited that he was really impossible to handle. I had come to recognize minute signals of this and always managed to find an excuse to shut him up in time, since it would have been disastrous to let him think he'd 'gaffered' me. I was therefore uncertain how he'd act before the TV cameras but decided to have a go.

The first thing was not to get him frightened. One of the packing-cases he used as a 'sett' to sleep in had an open top, the other a hole in the side. I fitted this latter with movable iron bars, so that I could shut him in his sleeping quarters if I wanted to, and I began to get him used to being shut in. For a few nights I left him there for about five minutes after feeding but always let him out if he seemed to be getting frantic, and I gradually increased the term of his imprisonment until he'd tolerate an hour or so by the time the great day arrived.

The performance was due at tea-time but I had been

asked to attend for rehearsals at eleven in the morning. And to make the act worthwhile, I took some guinea-pigs and ferrets to show the children as well. I borrowed a van and by the time Bill and his crate were loaded on, and a wire run for the guinea-pigs, a pen for the ferrets, travelling boxes, drinking bowls and straw, I looked like Noah with his Ark.

The producer was John Vernon, and it was my first appearance. I think it was almost his first show, though I didn't know that at the time. Anyhow, we were both desperately anxious to make it a success and neither of us appreciated the difficulties of the other.

First, he wanted to know exactly what I had to show. I had merely been booked to bring my badger and enough animals to make seven minutes. In addition to everything else I'd taken a hedgehog for good measure but we decided there was plenty without that and let him go.

The first time we went through the routine, all was well. The domestic stuff was easy and we decided that I should hold the badger up and talk about rearing and feeding him and then I should put him down and walk round in a circle about ten yards across, with him following at heel. It worked very well.

An hour later John asked me to do it again, so that the camera crew could see and get the idea of what they were to show. We went for a drink and some bread and cheese at the local, leaving poor Bill in his sleeping-box. After that he was a bit truculent – and I don't blame him – but they asked me to do it again with a path mown in the field as it was rather too long to show off his wonderful springy, bear-like action first time. I explained to John Vernon that you couldn't go on walking badgers round like lap-dogs indefinitely without their going on strike. He saw my point entirely but this was TV, the camera crews, stage manager and the whole bag of tricks must know exactly what was going on and I shouldn't bring animals I couldn't control.

He was the boss, so Bill and I did what we were asked, though I could foresee two snags. The only reason he followed closely to heel was that he was frightened of being lost on strange ground, and he was rapidly learning the territory; and he was getting fed up with being pushed around, so that I was acutely conscious that John wasn't going to be the only boss around there for long. To my horror, at about half past three, there was to be a final 'run through', to make certain everything was all right before the 'do'. I was becoming convinced that people on TV earned their corn. Bill was to be left till last because he was the obvious climax to my act. I got through the guinea-pigs and ferrets without a hitch and came to the spot where I had to open Bill's box and produce him in triumph, like a rabbit from a hat. Only he was a jolly big rabbit and his box was a jolly big hat.

I opened the lid and groped round for him, almost standing on tiptoe to reach the bottom. The moment I poked down I knew I was in for trouble for he grunted like a boar pig and tucked his head between his legs. He'd been out before; he was tired of my friends and he was comfortable where he was. Now when a badger arches his back and shoves his muzzle between his front legs, it is quite impossible to pick him up except by literally putting your hands exactly where his jaws are. Practically standing on your head, with the conviction that he means business if you do try to pick him up against his will, is scarcely calculated to increase the confidence of a novice in front of a television camera.

Fortunately, it could not record my facial expression. All that was visible was a vast expanse of backside. Furthermore, that was all that remained visible for some considerable time, because my calculations about Bill's frame of mind had been only too accurate. When I placed my hand under his belly to lift him out, he laid hold of it good and hard. He was not being vicious but firm. He calculated his hold so accurately that he held hard enough to retain me in my unnatural posture for what seemed like minute after minute, and yet when I came to examine my wounds later, he had scarcely broken the skin, though it had been agony at the time. When he did eventually relent, he went through the rest of his act perfectly, though I was even more glad that the sound microphone had been out of range than the camera.

Obviously a dress rehearsal like that did not help to control the butterflies in my tummy. But on the final transmission he behaved like the gentleman he really was and the children were captivated by his charm. For our part, John Vernon and I profited by the experience. When he appeared on the screen after that, we never gave him more than one rehearsal.

CHAPTER FIVE

THE STOATS

Strangers who came to see my animals often told me that I'd got a 'young zoo'. Any collection of animals and birds would be 'a zoo' to them and they expected me to be rather flattered by the remark. Nothing could be calculated to hurt my pride more. With the exception of a few wild duck, which have the full powers of flight and can leave me tomorrow if they wish, I never keep anything simply to be stared at like animals in the zoo. My interest in animals and birds has always been to get to know them and to establish a bond of mutual trust; to discover, for myself, where their environment makes them differ from their relatives in the wild, and to discover anything that I can which was not already known about the species. My concern, and indeed my excuse for keeping them, has been always to provide conditions for them which will allow them to be reasonably happy. So I am never very pleased about the idea that they are mere specimens to be inspected.

In a way, my stoats were an exception. I failed to tame them to my satisfaction, and, although my old dog stoat would submit to being stroked, it was nothing better than submission. I would be the last to pretend he really enjoyed it.

Bill Brock, the badger, survived his first winter and emerged into spring with a shining coat and immense appetite. While he had been growing up, Bert Gripton and I had got to know each other very well and worn the rough edges off our mutual suspicion. I'd discovered that the 'good tales' he could tell were not just stories, but that he was one of the finest naturalists I'd met. He, I think, realized that I was practical too and not only the theorist he'd imagined. It was inevitable, therefore, that we began to get along famously, because our tastes were so very similar.

One spring he sent me a postcard to see if I wanted a nest of stoats. He had been working in the fields and one of his dogs had marked them in a hedge bank and killed the bitch stoat. When he examined her, it was obvious that she was suckling young, so he dug out the hole and unearthed five stoat kittens, almost half-grown and well able to live on solid food. So he took them home and sent me word.

Unluckily, we were going away for a week's holiday and so I couldn't collect them till we came back. Bert fed them nightly on fresh raw rabbit and they were lively and healthy when I went for them.

It was a most critical period, though. Both Bert and I think that, had I taken them as soon as he found them there is some chance that I could have tamed them. It is my belief that it is possible to tame almost any young animal, with the possible exception of a wild cat, if it is caught young enough and handled often enough.

Handling was the trouble with these stoats, though. They grew in the week Bert had them, so that when I first saw them they were aggressive and large enough to make better men than me have second thoughts. Often as I've been bitten by various animals, I do draw the line at stoats once they are strong enough to get a really good grip.

Four of the five kittens were young dogs, exactly alike in every respect except that one had an odd white blaze on his muzzle. The fifth was a little bitch quite unlike any stoat I ever saw. I put them in an old budgerigar aviary, with a couple of small sleeping-boxes to choose from, and plenty of space to play over the wood-shaving-covered floor. The little bitch could easily be distinguished from her brothers because she was only about two-thirds the size. At first sight that was the only difference, and it was difficult to get more than a glimpse of them at first because they all dived for the cover of their sleeping-boxes the moment any human being got near. Then we noticed that the bitch was just a trifle slower off the mark and a little less agile at getting into the hole of the sleeping-box. Perhaps she was a bit off-colour.

When we discovered the real reason, we just didn't

believe our eyes. Both her back legs had been amputated, the one above the foot and the other right high up in the thigh. I caught her then to examine her properly, as I thought she had been fighting and that it would probably be kindest to destroy her. But both stumps were completely healed and covered with a horny callus of leathery skin which allowed her to shuffle along far faster than lots of animals can run, with no sign of discomfort at all.

I think that if there had been another bitch in the litter I should have killed my cripple to have put her 'out of her misery'. As it was, I persuaded myself that she was perfectly happy and kept her! As things turned out, I think I was right. No stranger ever realized she was deformed until it was pointed out.

Lots of people have advanced theories about the origin of her injuries and I have come to the conclusion that either the bitch stoat was carrying her from one hole to another and sprung a rabbit trap, which missed the old bitch but caught the kitten by the back legs, or she had trouble when it was born and bit off its legs herself to help parturition. I think that a stoat finding her kitten fast by the legs, whimpering in a rabbit trap, would bite it free in her frenzy; and I think that the elementary surgery of such crude amputation is perfectly feasible in case of real difficulty at birth. But I shall never know for certain what is the real story.

Having got my stoats installed in their aviary, the first problem was to feed them, and the next to tame them. Feeding in May was no trouble at all because, at that time, I lived at Bloxwich and my chicken-runs were infested with sparrows, which flocked from the surrounding streets to my fowl corn. So a sparrow trap, left permanently set, provided all the young stoats could eat. The danger in not varying the diet, however, is that when it eventually becomes unavoidable to supply something else, it is often impossible to persuade the animal concerned to eat the alternative. I got a constant supply of fresh fowl-heads from a friendly poulterer, for example, and for some weeks fed a weasel exclusively on the freshest heads. Then the supply temporarily ceased and I tried ordinary raw meat, and even guinea-fowl heads, with no success at all. He would eat nothing, except fowl-heads, until literally starved into it.

When I read of the damage that badgers, foxes, or rooks are doing, I often wonder about this intense craving for one specific food. I have found that, with captive animals, it is quite impossible to generalize about diet. They often take to the most surprising foods, almost to the exclusion of all else, to such an extent that they would rather go really hungry than accept what appears to be a natural alternative. Allowing for the dangers of judging wild animals' behaviour by captive standards, it does seem to me likely that the same may happen in natural surroundings. Because an odd badger, for instance, discovers that lambs are good to eat it is quite sensible to kill him. But it would be unreasonable to assume, from that, that all other badgers have similar tastes.

Anyhow, I was very careful with my stoats. Right from the start I gave variety, with never quite enough of one dish to satisfy them, so that they always had to start something else. Then I discovered that one of their favourite dishes was raw egg. I've always got ducks and poultry so, for some time, raw egg became their staple diet, supplemented by whatever fresh meat I could procure. And they had milk and water in separate dishes.

The real problem, however, was how to tame them.

I have already said they were quite big enough to give shrewder bites than I relished, and they could strike as fast as rattlesnakes. Every time I disturbed their sleeping-box there was a chorus of high-pitched chattering, little staccato yelps of fury. I remembered the days of Hairy Kelly, when I dared not put my hand in a sackful of rats, and twenty or thirty years had given me no higher courage.

I decided to get them used to me from a distance. The aviary where they lived was on the edge of a little lawn, so I started to feed them at night exclusively on raw egg. Within two minutes of putting it in their pen, one cautious, brown, snake-like head after another popped up out of the sleeping-box to look round. By then I had retreated five or ten yards up the lawn – depending on whether the wind was towards me or them – and 'frozen' as still as a tree.

One minor trick of self-discipline that I have acquired over the years, is almost an indifference to flies. They can crawl all over my face or neck without making me bat an eyelid, and I have found that, once I acquired this knack of not minding them very much, it is less trouble to endure them than all the hand-waving and arm-swinging I used to indulge in.

This trick of keeping still, even in uncomfortable positions, served me well with my stoats. First one and then the next would slide out of the box and glide smoothly round the aviary keeping close to the walls. At the least sudden sound or movement four furry dotted lines traced the quickest and most direct path back 'home'. And then, within a minute, the whole pantomime would start all over again. While all this was going on, the little bitch would stay under cover listening, always listening, for the first hint that her brothers were eating. Then, in a twinkling, she would be snarling and spitting in the centre of them, fighting for the biggest and juiciest piece.

The eggs I fed them on at this time were juicy enough, because they were raw, but a great deal of them went to waste. I simply used to crack the egg and put it on the floor of the aviary, which was thickly covered with wood shavings to make a soft playground for the little cripple. First one young dog and then another would pluck up courage to get close to it. They would nuzzle it over, searching for the crack, and then they would discover that it was too large and smooth to bite properly. They couldn't puncture the shell. The more they nuzzled and struggled, the faster the egg rolled away from them. Sometimes they'd be lucky and it would smack into the wall or doorpost almost at once and they'd all begin to feed. But quite often, they'd all begin shoving at once from every side, like forwards in a rugby scrum. And to me, their spectator, this little scrum of stoats, all heaving and struggling in opposite directions, so that the 'ball' hardly moved, despite their efforts, it seemed like far-off boyhood days. I would stand there on the lawn, far into the summer evenings, and relive those sweaty, smelly struggles, when nothing mattered but to beat the other side at getting an oval rugby ball out of a sticky mound of bodies.

The stoats had more sense than we had, though. At least they could eat the 'ball' they got out of their scrum. Their routine was always the same. The four dogs would chatter and struggle for possession of the egg, without either side showing superiority. But I was not the only spectator. From the safety of her box, the bitch would peer out until the exact psychological moment when she calculated one side or the other was flagging. Then she would slide out and hover for a moment on the edge of the scrum before diving, more smoothly than an international scrum half, to emerge shoving the egg in front of her. At once the rest of the pack would rally round like forwards dribbling for the touchline, and there would be the inevitable smack as the shell cracked on the wall. One second there was shuffling, chattering pandemonium and the next the stillness of a summer night.

I enjoyed the game as much as they did and, in my own way, I was just as active. Half the art of stalking in the wild is to slide noiselessly forward, a step at a time, whenever the quarry is distracted for an instant. You've only got to watch a cat stalking a bird to prove that. I couldn't move much till the bitch had joined in the fray because, although it was fairly easy to approach the dog stoats, whilst they were busy with the egg, she wouldn't come out if she saw me. So I used to make the most of the few seconds after she joined in, until the egg smacked into something solid. I would drift one, two, three, four, sometimes five paces forward, until I was almost up to the netting of the aviary, and could watch whatever went on at arm's length. If I was caught moving up, there was a mad rush for cover and that was the last I would see for that night. I'd lost the game.

It always puzzled me to find out why I could walk openly to the pen, make quite a clatter about putting the egg in, and the whole litter would be out and playing within two minutes of the time I went away. Yet if I went back either obviously or by stealth, they would dive for cover and refuse to come out. It was just possible to fool them, though. If I was in a hurry to see them, I would clatter loudly to the pen, make much ado about putting the food in, and clatter loudly away again. With all that noise they would cower deeply in their bedding for cover and, before the echoes of my departure had died away, I would drift back right up to the door itself. After the usual two minutes' silence, one after the other would slide out and carry on as if all was safety and solitude.

After a week or so they got used to the smell of a human and I was able to take my stalking a step further. The aviary had a partition which I shut, so that I could open one door to the garden and begin a series of stalks ending right inside. Success begot success, until I could shut myself in, open up the partition and eventually they would ignore me as if I was a statue, and roll and gambol and chatter all round me, even over my feet.

Then I made a discovery which thrilled me beyond measure.

I had been watching them play with their egg one night, chattering loudly in excitement until it broke and they all began to lap it up, before it became mixed too inextricably with the wood shavings on the floor. I was surprised, in the silence, to find how noisily they lapped and smacked their chops about it, and swore at each other beneath their breath.

Then they began to play their 'leaf game', which was really very simple. Each, in turn, would assume the role of leader and start rushing round in circles and figures of eight, like a puppy having a 'mad fit'. But the others would tag on behind, literally nose-to-tail and the whole litter would whisk round for all the world like leaves in a whirlwind. They went not only round the floor but up a sack-covered ramp, round a shelf and down the other ramp, so fast that it was literally impossible to count them.

Twice, in the wild state, I have seen a stoat doing this traditional 'dance', to the complete bewilderment of birds which saw it. But never, except with my litter, have I seen a communal stoat-dance. There is a theory that stoats use this mad, whirling ballet to arouse the curiosity of their prey, so that they can approach near enough to strike, like a snake, from under the cloak of their artistry. If that is true, and I believe it is, then this dance performed by my young stoats was merely preparation for the future. Almost all animals' play, from the mock battle of puppies, to young hares jumping and twisting into the air for no apparent reason, is connected with the learning of a completely automatic routine for attack or escape or defence. My stoats, I think, were learning their lessons communally, and they certainly burnt into my memory a picture of rhythm, speed and grace which will give me pleasure for the rest of my life.

When they had finished their play, for a moment, and were exhausted, they would settle down to their toilet, licking and preening every disarranged hair into place as carefully as a cat. Like cats, they would often leave off washing themselves to attend to the other members of the family and the whole time they were caressing each other they would croon with affection and delight.

It was a tiny, musical, high-pitched sound, with about the frequency of the purr of a cat. I have never heard or read of it, partly, I imagine, because it is unusual to be able to get near enough to more than one stoat without being noticed. But I used to sit out there in that aviary, when my neighbours were in bed, enjoying my role of unnoticed and uninvited guest at a stoats' supper-party.

I spent hour upon hour with a microphone trying to record the affection call of Bill Brock, my badger. When I first went to see him in the evening, he always reared up the half door to his loose-box and made a sound which I can only describe as a very faint whinny, like a stallion does. It was an oddly confidential call, obviously designed to express affection for one person only. And it was almost exactly like a louder, bigger, rather slower edition of the affection calls of my stoats.

Naturally the discovery enthralled me. Somewhere, in the back of my memory, it was a song I'd heard before. Then I remembered. Most country folk have kept ferrets and will have noticed the rather high-pitched whinnies when two are introduced to mate and the jilt ferret is in season. I have rarely heard them do it at any other time, but just when they are preparing to mate, before the hob ferret dispels all sense of romance by his brutal, caveman courtship. Just for their few, brief, rose-tinted moments, ferrets too make almost exactly this same noise.

I cast round for the connecting link. Stoats and badgers, polecats and ferrets all belong to the Mustelidae, or animals which carry stink glands. Badgers don't look very much like the others, but at least they have the stink gland in common.

So do otters and weasels. I have never yet had an otter, though I hope one day I shall, but I have had three weasels, one of which was so affectionate that he would give this wonderful crooning song of welcome every time I troubled to tickle his belly with my finger. And he, too, made almost exactly the same sound. Not, of course, in the same key or with the same volume, but quite obviously an extremely similar call. This one was easier. I got a recording of him for the BBC that was absolutely perfect. And if I didn't know the Mustelidae were related by physical similarities, there would have been no doubt whatever in my mind when I heard their voices.

There is a by-product of joy to be had from keeping animals and birds. They are a source of introduction to some of the nicest people.

One of my heroes (or heroines) when I was at school, was the naturalist and writer, Miss Frances Pitt. She was a regular contributor to one of the great London evening papers from the 1920s until she died in 1961 and I have read her books about her tame otters, badgers, ravens and squirrels and many other British animals and birds, time and time again. But if I had not later begun to follow her example myself, it is unlikely that we should have met.

When my stoats were about nine months old, it became obvious that I couldn't keep them all. The dogs were beginning to be rather churlish with each other and, in any case, the bitch would have had rather a thin time with four such suitors to contend with. For any man to be known as 'a bit of a stoat' is no idle insult. I decided, therefore, to keep my bitch and the dog with the white markings on his muzzle, for he was the tamest of all as well as being distinctive. The problem was to dispose of the other three. I could easily have turned them loose in some rabbit-infested spot, in those days, but, if they were not tame enough to handle, I had spent a great deal of time with them and they weren't bad, for stoats. So the next time Miss Pitt came over, I asked her if she would like them. She accepted two and I was very curious to know what she intended to do with them. With her usual hospitality, she suggested we take them over for lunch the following Sunday so that we could see for ourselves.

One of the problems about keeping wild things in unnatural surroundings is that all sorts of difficulties arise when it becomes necessary to move them. I always take special precautions, with anything I know I shall have to move much, by having a sleeping-box, or a small cage, within its cage, which is easily transportable. Wherever it goes then, it always has a refuge it knows; its surroundings are never entirely strange. I hadn't anticipated moving my stoats and there was nothing easily portable in their aviary except the sleeping-boxes, and they could not readily be shut up.

I decided a cage rat-trap would meet the bill since I could get to Miss Pitt in half an hour, so any travelling cage would only be very temporary. First I separated the bitch and dog I was keeping, persuading them into one sleeping-box. It was then easy to guide a couple of the others into the funnel, from which they could not come back. Everything went according to plan and I got them into my wire cage with no flurry or panic. They naturally rushed round a bit, in a vain attempt to escape, but I was horrified when one rolled over on its side and began to twitch. If he had had a long and arduous chase, or been handled, or otherwise terrified, I should have put it down to my own mishandling. As it was, I could see no real cause for such panic, though I was very relieved to see him 'come round' within about thirty seconds and appear to be none the worse, except for a rather faraway look in his eyes, like a puppy that has just had a teething fit.

An hour later they were both installed in their new quarters at Miss Pitt's home. She had got a roomy shed wired in for them, entirely devoid of cover or sleeping quarters on the ground. At one side there was a rather complicated sleeping-box, raised from the floor, with a ramp each end leading up to an entrance hole. This box was divided into two parts by a sliding wire-netting partition, which could either be left in place, so that access to either end could only be made from the correct ramp, or it could be removed to make one large box with an entrance either end.

I was very curious and, when Miss Pitt said she wanted to discover if it were possible to cross a stoat with a ferret, I was very sceptical as well. She had already got a jill ferret used to the box and its surrounding shed, and she had fastened it in the sleeping-box, to give the stoats the opportunity of learning the runs of their new territory.

As a child I had a favourite and much petted ferret, which was so bitten by a stoat that it had died. And I had had several ferrets since, quite able and willing to give as good as they got. So I tried to forget any personal affection I had developed in months of acquaintance with these two stoats, and persuaded myself to take a detached and scientific interest in this experiment. Happy hours reading Miss Pitt's books should have taught me better. She wouldn't have started the experiment if she hadn't been confident that, at least, the result wouldn't have been a bloody battle.

The stoats cringed a little, and searched round and round the floor, in a vain attempt to find a way of escape. Within half an hour they had climbed the ramp into the sleeping-box and found the only available cover. 'Cover' was true only as far as vanishing from sight was concerned. When they were hiding in the box they could still see and hear and smell the ferret a few inches away, with only wire-netting as a barrier.

I was so enthralled that I wanted to try the experiment for myself and, with characteristic generosity, Miss Pitt expressed her interest in getting a cross check, although I was shamelessly pinching her idea.

It was easy in my aviary. It was already divided in two, with a wire-netting partition, and all I had to do was to fit a divided sleeping-box, with one half opening into each compartment. I had an old jill ferret, who had been a simply fearless ratter for three years and had then been retired on her laurels. When ferrets have suffered a varying amount of punishment from rats, they cease to be game. This old lady would sail into a rat hole with all the aplomb in the world. But if she found an old rat cornered, and prepared to fight it out, she would turn her head slightly, as if she just couldn't see it, and remember what a pressing engagement she had elsewhere. I always liked a ferret like that. You didn't catch all the rats but, at least, the first half-hour after getting home could be spent in eating, instead of dressing the wounds of ferrets with more pluck than sense.

I was pretty sure, therefore, that my old dear wouldn't want to go to the extent of doing battle with the stoats, even if she wasn't friendly.

Neither liked the smell of the other very much, so that each made up a bed at opposite corners of the sleeping-box. I had, of course, confined the experiment to the ferret and one dog stoat. The little bitch and my chap with the white blaze were moved to other quarters.

One Sunday, while I was at home all day, I lifted the partition. The old ferret was just coming into season, so she was as attractive as she would ever be. There was nothing more exciting to distract the dog stoat's attention. At least there was no unseemly brawling. Neither was there any sign of mutual attraction. Each remained in the accustomed corner of the sleeping-box; each used the accustomed ramp and the accustomed half of the aviary; each appeared to be the owner of an established territory and neither ever crossed the imaginary dividing line.

It was a superb example to us humans of peaceful co-existence. Here a ferret and a stoat, traditional enemies, were living in toleration within the confines of a small aviary and sharing an even tinier sleeping-box. But neither ever forgot the boundaries of propriety. They shared the same raw meat and drinking vessels from sheer necessity. They never forgot themselves so far as to share the same bed.

I allowed the experiment to continue for several weeks, with no result at all; but it is dangerous to ferrets to remain unmated when they come in season. They often die of complaints apparently peculiar to females of their species. So I gave the dog stoat, who had lived celibate so long with her, to my friend Donald Risdon of Dudley Zoo and mated the ferret to a polecat. Miss Pitt's experiment was equally non-productive, though hers was brought to an accidental end by the escape of both her stoats through a hole they enlarged from a mousehole. If I ever get another stoat young enough, I shall rear it with a ferret of the opposite sex because, though I think it unlikely that a successful mating would result, it is certainly an experiment worthwhile persevering with.

I kept the remaining pair of stoats but, despite quiet and every opportunity, they never produced a litter. Whether this was because they never mated, or whether they had young but destroyed them themselves, I never found out.* It was impossible to know for certain when the bitch had been in season because I would take no risks of handling her or disturbing her unnecessarily. But I kept a very sharp eye on her figure, without ever detecting the obvious signs I should expect. Nevertheless, she remained perfectly healthy, so it would appear that stoats do not suffer the ill-effects of ferrets if they do not breed.

I did, however, get one odd clue. The dog stoat seemed very seedy during his second breeding season and then, after a period of lassitude, I noticed he was losing the sight of one eye, which turned a sickening sightless blue. I often tried to conjecture whether the cause was some accidental collision during his acrobatic capers or whether his bitch had given him a bite to make him remember his manners.

For a cripple, her powers of self-preservation were nothing short of miraculous. I went to clean out their aviary, during the summer, and looked carefully round to make sure they were in the sleeping-box and that it was safe to open the door. It did not occur to me to look up as well as down. So when I opened the aviary door, I was horrified to see the little bitch clinging to the top of the wire-netting. Stealthily, I tried to shut her back in but, already, she was half-way out and I couldn't close the door without squashing her. I eased it for a moment but, instead of going back in, she squeezed out and on to the roof.

Now, a lifetime persuading animals to go where I want should have made me perform the correct routine as naturally as a frightened rat dives for cover. My poor, scared little bitch was cowering on the roof, trying to make up her mind which side to jump off, but lying as still as a

*See author's note p. 108

twig whilst she calculated. Four feet away was a bit of iron spouting the litter had used as a plaything, diving through it after their leader, like trucks follow a train through a tunnel. All I had to do was to put this pipe slowly, and smoothly, on to the roof and my stoat would have scurried into it for cover, as she had done in play so often as a youngster.

My mind was just like porridge. Instead of acting smartly and quietly, to give her something to hide in, I rushed to the house for my wife. When she came out, I went to the front of the roof and she to the rear in an abortive effort to drive the bitch stoat back through the open door. It was like trying to bottle a sunbeam. In spite of her crippled back legs, she gave a dive into space, from the roof, right over the wall into my neighbour's garden.

We all trooped round and spied her taking cover under a great clump of lupins. Even now my addled mind hadn't grasped the fundamental I'd been learning all my life. Instead of putting a pipe or familiar box down for her to use as refuge, I made one mad, heroic, headlong dive to try and grab her in my hands. She was so precious to me that I would have hung on, if I'd caught her, whatever the cost in bites might be. But even a crippled stoat is more than a match for me, both physically and mentally. While I was still poised in mid-air, she had braved the space across the lawn and taken refuge in a huge impregnable pile of bricks.

My neighbours had young children and feared that they might be quicker than I and get bitten for their trouble. A seven-foot wall separated the stoat from her mate and her aviary, so I was sure she wouldn't get back of her own accord. I couldn't think of any way of catching her alive, and I was obviously going to be unpopular for letting a dangerous beast escape, so long as she remained at liberty.

Sadly I went home and fetched the dogs. A hunt for a wild stoat would have been great fun; chasing a healthy one that had escaped, to prevent damage, could have been an interesting exercise of skill; to try to hound my beloved cripple to death, simply because I had been dull-witted enough to muff my chances of getting her unhurt, was bitter medicine indeed. I needn't have worried. She easily eluded the dogs and me in her new-found heap of bricks.

I baited rat cages with sparrows and put sleeping-boxes and saucers of milk and water in strategic places. For a night or two she came for a drink and burrowed under the wire cages to feed, in safety, on the sparrows through the mesh of the cage. Then she disappeared and I gradually gave up hope.

I was lucky that it was summer weather, for I usually took the dogs for a walk just as the light was failing. Ten days later I was at my father's home, which was across the main road, behind the houses opposite ours. I had crossed his field and was leaning on the gate, looking idly into his garden, when I was suddenly galvanized into utter im-mobility. There, slipping along down the hedge-bottom, as if she owned the place, was my little bitch stoat. I'm a confirmed leaner on almost every gate I come to, because I know of no other way of keeping perfectly still so easily. This time I was successful. She didn't see me and slipped off down some rat-holes by the fowl-pen. It was the time of year that never really gets dark, but I stayed where I was till my eyes were nearly squeezed out in an effort to probe the gloom for her. That one glimpse was all I saw.

Eventually I forced myself away and went home to fetch

the rat cage and baits and all the paraphernalia I'd concocted. Next morning I was up before it was light, without any result at all.

Two days later I noticed that a nest of young pigeons, housed in a pen I kept in my father's field, had disappeared. I was rather annoyed because I had had a kit of Oriental Rolling Pigeons, which had delighted me with their aerobatics for years. It is particularly hard to breed good ones; either they are too good, and somersault down in a death-crash, or they hardly roll at all. This young pair that I lost were by the best cock I'd got. The next day another went and I suspected someone's cat. They just disappeared into thin air. I moved everything round before coming across them, all three, at the back of a nest box. Rats were the obvious culprits but rats usually split the crops and eat any half-digested corn there is in them. These had been bitten about the neck, back and eyes. I wondered about my stoat.

I took the dead pigeons away and baited my rat cage. Nothing happened for two days and then another pigeon went. This time I was sure it was my stoat. I could have caught her easily in a rabbit gin and saved my pigeons. But I could breed more pigeons far easier than I could get another stoat, so I bore myself with patience. By the time seventeen days had elapsed since she escaped, six pigeons had paid the price, though she was obviously killing more for fun than food. She ate very little of them. I realized she was probably feeding mainly on field voles or young rats, and that she might move on as soon as she had skimmed the cream off the local rat and mouse population. I was frightened of sacrificing a perfectly good kit of pigeons and still losing the stoat in the end.

The Stoats

Then I remembered an old keeper's trick. Stoats are particularly curious about other stoats, and keepers often empty the urine of a shot or trapped stoat on to a stick and set a trap by it. Other stoats call, like dogs to a lamp-post, and spring the trap.

So I bolted my dog stoat at home into the rat cage, which I had put on a sack, and I chivvied him round so that, in his panic, he wet it pretty thoroughly. I let him out, back into his aviary, and took the trap and sack over to the pigeon pen, from which I removed all the remaining pigeons, so as to leave nothing to distract the bitch if she came.

Next morning, to my delight, she was safely in my trap. Ever since she was a kitten, she'd lived a completely sheltered life with her brothers in my aviary. It had never been necessary for her to exert herself to catch her food, and she had never even seen live prey. Great care had been taken not to frighten her, so that there had been nothing on which she could build her traditional cunning and fear of man. She was such a cripple, with only two good legs, that it seemed quite impossible to me that she should survive at all at liberty.

Yet here she was, chattering brave defiance at me, after two and a half weeks at liberty. She had killed seven of my best pigeons, and avoided my traps with all the artistry of her kind. Her coat was a little dull and there was an enormous tick, as large as a sheep tick, between her shoulders. That, I suppose, was because she couldn't scratch with her back feet.

I put her gently back in her aviary, and stood quietly by the door to see what happened. The joy of her mate was pathetic to see. Litter-sister or no litter-sister, his relationship was obviously more than brotherly. He licked her from head to foot, to erase all the foreign smells she had acquired; he crooned more love songs than Bing Crosby, and my fingers itched for a microphone to record his rhapsody; next day the tick on her shoulder had gone, so he must have performed her toilet for her. The bitch, on the other hand, was utterly undemonstrative. She took all the fuss and affection for granted, like any other spoilt woman.

But she was content to stay at home after that, and I don't think that this was entirely due to the extra precautions I took since she escaped.

Author's Note

I have bred stoats in captivity twice since that. The bitch stoat mates in summer but, although the ovule is fertilised, implantation is delayed and it floats free in the uterus until the following spring. Only then is it implanted in the uterus wall, after which a normal gestation of about eight weeks takes place.

This delayed implantation was not properly understood when I had my first stoats but it is now known that male and female come together only for mating. The fact that I housed my original pair together could well have meant that, even if the bitch stoat did conceive, antipathy between them would have been such that any youngsters born would be destroyed.

The bitch stoats have been dominant in the pairs I have kept and I now doubt if the original pair would have shared the same nest box if they had had any choice.

CHAPTER SIX

BIRDS

Most of the birds I have kept have been but temporary guests. At school I had some very good jackdaws, which I used to take home in the holidays and keep more or less at liberty. For some weeks they came down to be fed but, gradually, as they learned to fend for themselves, we drifted apart until they were entirely self-supporting.

That is one of the few advantages which birds have over animals. It is usually possible to liberate them gradually, in the knowledge that they will suffer no great hardship. Animals are very different. They often lose their fear of man and raid his property, as my stoat did, or they fall easy prey to some marauder, because they have never had to learn to look after themselves.

It is sometimes very difficult to get rid of tame birds satisfactorily, though. Some years ago, a young tawny owl arrived at my house because it had been rescued from a gang of young thugs who had been stoning it. It had just left the nest and learned to fly, but one of these children had evidently hit it a tremendous blow, because there was a gaping wound on the butt of one wing, which was about as large as an old penny. My first reaction was to pull its neck and make the best of a bad job by giving it to the ferrets. But, when I examined it closely, it was obviously only a surface wound, fresh at that, and there was no shattered bone, as I had feared. So I filled the cut with M & B powder and popped the young owl in the dark woodshed to recover his composure. That night I took him a couple of sparrows, from my never-failing trap, but he didn't seem to know what to do with them. As I approached, he hissed and snapped his beak at me in a most ferocious way, but he obviously hadn't much idea of feeding himself. I was surprised at this, because birds of prey usually drop food on the nest and their young learn to tear at it from a very early age.

Next morning, there were a couple of 'pellets' on the shelf where he was roosting, though the sparrows weren't touched. Now owls, like other birds of prey, habitually eat fur, feathers, bones and all when they have a meal. Indeed, it is necessary to their rather peculiar digestions that they should. Their digestive system absorbs the nourishment that they have eaten, and compresses the indigestible bulk into pellets which they vomit back.

I took these two pellets that my young tawny had ejected, and carefully unwrapped them. They were each about the size of a cigar butt, slightly pointed at each end, brown, tightly packed and fibrous. I shredded them carefully in my fingers, and the first thing which became obvious was that they contained a great deal of short brown fur. Mice or voles of some sort. The conjecture was strengthened by the presence of 'long bones', which had come from the front and back legs of mice, being far too short for leg or wing bones of birds, and also the shoulder-blades. I wanted a skull for confirmation. The second pellet I shredded contained a great long skull with the yellowing fangs of a field vole.

That faced me with the problem of catching mice for my owl or changing his diet. Whenever I come across a roost of either hawks or owls, I always examine their 'casts' – as these pellets are called – partly to satisfy myself about their diet, and partly as an accurate way of deciding what the local small bird or mammal population consists of. If, for example, there is a plague of voles, shrews or dorbeetles, the fact is usually obvious from the contents of these casts. If there is merely a wide selection of species, it will be easy to see a cross section of it – in both senses.

I felt under no great obligation to my tawny owl. If I had stolen him from his nest I should have felt in honour bound to cause him an absolute minimum of discomfort. As it was, the boot was on the other foot. For once I was simply doing him a good turn by trying to cure him of the ills caused by children who should have known better.

I wondered if it would be as difficult to get him to change

his diet as if he had been one of the animals I kept. And I was a little apprehensive about his digestion. The pellets he had cast up were from food he had eaten before he was stoned and I would have expected him to have cast them before, instead of retaining them until he came to me. Perhaps his injuries were internal as well as superficial.

I needn't have worried. The next night, I broke one of his sparrow corpses in two, so that the flesh and bone were exposed, and stood quietly in the shadows to watch. He fell upon it like a tiger. Standing on it with his powerful talons, he slid his vicious hooked beak into flesh, feather and sinew without distinction. It came away in neat shreds, as easily as if it had been corned beef. The first night, if I approached, he cowered into a corner. After that he stood his ground, his eyes flashing defiance, like amber beacons in the road, and his beak clacking as if he would shred me limb from limb, too, unless I was civil.

He was very intelligent. Within four days he had learned that I was the bringer of food, and would step delicately on to my finger like a falcon at fist. When he had finished his food, he would stay perched on my hand or shoulder, preening himself and turning his head round on his shoulders, in that absurd way that owls affect. I used to marvel at the delicacy with which he could control his powerful hooked beak. When he came he was scarcely past the downy stage. His feathers were still dull and their bases surrounded with the membranous, rather scurfy envelope which guards them from harm as they grow. As the skeleton of the feathers hardened, he would take each in turn gently in his beak and lift away the membrane with astonishing precision.

I had one terrifying personal example of this dexterity. He was sitting on my shoulder performing his toilet, until there wasn't a single feather out of place. Then he noticed the tiny hairs on the lobe of my ear and, before I could do anything to stop him, he had taken it in his beak to groom it to his liking. The hair rose on the nape of my neck as it dawned on me what he was doing. I couldn't be certain whether he was paying me the compliment of tidying me up a bit, or whether he had suddenly noticed the lobe of my ear and was testing it to discover if it were worth eating. Slowly, gently, imperceptibly, I slid my hand under his breast and raised it. He stepped obediently on to my finger and I moved him away from my head. Without pausing, he started to give attention to the hairs on the back of my wrist. We had obviously become firm friends. This time I had the advantage of being able to watch and admire his skill.

For the next week or so his appetite was enormous and he would regularly consume either five or six sparrows in a night. It became necessary for him to have flying practice, so that his performance in the air should match his size. The little woodshed where he lived opened into the verandah, so, each evening, I opened the door and spread his sparrows about so that he had to hunt over the whole area for them. He would come out and perch on the open door of his shed, while he surveyed the wider scene, strewn with the corpses of half a dozen sparrows. Then he would open his wings and float down on to one, like a grey-brown ghost. However close I stood, there was no sound from his feathers, which were quite perfectly designed to allow silent flight. And I wondered how many ghost stories owed their origin to some old tawny owl, flying close enough for a lonely soul, in some quiet lane, to feel the wind and see a dark shadow – but hear nothing at all.

The combination of food in plenty and freedom to fly a little at night, soon worked the magical transformation from a helpless owlet to a self-confident owl. By then, I had become very fond of him, but corn harvest was approaching. And when the corn is ripe, the town sparrows migrate to the fields in their thousands for their summer holiday. I knew from experience that it would then be quite impossible to provide six sparrows a night.

The obvious thing was to let my owl go. His wing was healed; he was plump enough to stand hardship when he first found that freedom did not necessarily mean plenty; he was strong on the wing. So one Friday night I went across the road and let him go in my father's field. He sat on my arm and looked round until I thought he would screw his head off. Then he opened his wings and floated majestically up into a big elm tree, and never so much as looked back to give thanks.

The next evening I took some sparrows and placed them for him in the centre of the lawn, where he could see them from most of the trees in the garden. I could hear him up a tree making the odd wheezing call of young owls, but he took not the slightest notice of me or my sparrows. The same thing happened every night for a week, his calls becoming more insistent and pressing as time went on. But he still neither came to me nor the food I put him.

I came to the conclusion that he was catching his own food and that it was a waste of effort on my part catching sparrows for him. On the tenth night, when I went across, he was waiting on the centre of the lawn. I threw him a sparrow and he toppled clean over when he went to pluck it. He was literally too weak with hunger to stand up.

I picked him up and took him tenderly back home. Like

so many creatures which have known captivity, he was quite incapable of being self-supporting in his liberty. In most cases, animals or birds just creep off to die in obscurity; my owl was luckier. He had stayed about, near where he found freedom, long enough for me to rescue him and return him to his security.

I still had no intention of keeping him longer than necessary. At first he was so weak from starvation that I had to break his food up into pieces small enough for him to swallow. But, within a week, he was again as strong as ever and I intensified his training in self-reliance.

Next time I liberated him there was no mistake. He began where he had left off before by coming down to the lawn for food. When it became obvious that he was just getting lazy, and relying on me as his universal provider, I cut down the quantity and replaced sparrows by raw meat. That worked marvels. He soon became a good enough hunter to be able to ignore my charity.

Although he had been so affectionate when I had been hand-rearing him, he never showed any sign of recognition after he learned to look after himself. But, as luck would have it, the area where I let him go was pretty free from wild owls. It was on the edge of a large industrial belt and I suppose that the wild owl population found it easier and pleasanter to frequent the land to the north, which contained fewer houses and buildings. So my young owl took to our fields as his own territory. Next spring he had also acquired a mate, and either they or their descendants have been there ever since.

In total, I suppose, I kept him no more than four or five weeks, and he was one of the very few birds or beasts which have accepted my hospitality without teaching me anything special in return. Yet, for some reason, I was very fond of him and decided, if possible, to repeat the experiment with barn owls.

Somewhere or other I remember reading that barn owls never roost or nest in buildings below about fourteen feet from the ground. I've no idea at all whether that is true or not and, from my personal experience, I would doubt it. I find barn owls one of the prettiest and most attractive of our native birds – examine one closely if you've never thought of them in that light – and I have always wanted a pair living naturally about my house. They are, of course, on the protected list, but I'm hoping a couple of waifs might come my way. I would rear them in a small loft which I hope they would accept as home. Then, as soon as they became self-supporting, I should leave them to fend for themselves, in the hope that dusk round my house, from then on, would be enlivened by drifting white ghosts who had once been my friends.

One of the reasons we bought our last house was that there was a rookery in some tall lime trees, just outside the kitchen window. This rookery, and a stream running through a pool, were about the only attractions. The house itself was so dilapidated that it took five years of our energy to make the place habitable. But we didn't buy it for the house. There was a field with old trees and shrubs for cover; while we ate we could look out at our ornamental waterfowl and the wild birds which came to nest on the far side of their pool, which we had fox-proofed. There were outbuildings, which got first priority in our reconstructions, because that was where the animals lived; and we woke in the spring to a chorus of rooks.

It is unfashionable to be attracted by rooks. Farming journalists, gamekeepers and pseudo-scientists all revile them. But on balance I expect they do about as much good as harm and, for me, they breathe the very spirit of English villages. Against a common foe they unite, and curse him until he shrivels like a bureaucrat who has suggested building a bus shelter on the village green. A flock of rooks will probe over a pasture for leather-jackets in complete unity. Yet when they get home to the rookery, there is as much squabbling and bickering as at an annual garden-party. Above all, the noise of rooks is as much associated with country things for me as woodpeckers or woodpigeons deep in a wood. Church bells and rookeries are often found in towns as well. But both always seem slightly out of place.

I was discussing this question of rooks with a journalist friend one day and he told me that in his part of the country, in Essex, there had been a run of white rooks. There is a rookery at the rectory in Ashingdon, where an albino rook was hatched, which was taken to the London Zoo where it lived for several years. My friend, John May, said that if there was one the next year, he would try to get it for me. I was happy enough at the prospect, because I didn't give much for the chances of an albino rook left wild to its own resources. There are too many fools with guns who will blaze at anything unusual. Anyhow, I didn't think it particularly likely that there would be another albino or that John would be able to get it for me if there was. I was keen enough to remind him about it in my Christmas card, though.

My birthday is in the middle of May, and I usually have an evening rook-shooting at about that time. Fond as I am of rooks, I believe that any surplus in their numbers is better in rook pie than at large, getting the rest of their species a bad name. So about 10 May I began to wonder if John was going to contact me, or if the promise of last summer had died in the bar-parlour where it was born.

On 12 May, I received a telephone call to say that 'this year's white rook' was waiting collection at Ashingdon, Essex. It was a matter of urgency and trains were out of the question, so he came by road and arrived at eleven that night. Although he had only been caught the day before, there was no difficulty about persuading him to feed. He had even fed twice in the car on the way up.

I was simply delighted. He was a complete albino, with pink eyes, white legs and bill, and feathers outshining the purity of the famous washing-powder advertisement. By nature, rooks are shy and suspicious birds. They would be exterminated if they weren't. Not so my young rook. The obvious name for him was Snowy and, right from the start, he seemed naturally tame.

We fed him like the jackdaws of schooldays. Bread and milk, hard-boiled egg, bits of meat and pigeon grit. Because he was such a rarity, I bought ten bob's worth of mealworms too, so that he should have live food. There was a certain amount of argument about the grit because one school of thought held that rooks didn't need it – certainly not fledgelings. But anyone who goes out early in the morning by car, cannot fail to notice almost as many rooks as pigeons 'gritting' along the roadside. So I backed my hunch and dipped a little food in grit each day, to make sure he got some.

All went well for a while but the mealworms seemed unduly expensive, and I replaced them with 'gentles', the bluebottle maggots that fishermen use. Only I didn't buy them, I bred them – by hanging a horse's lights up and allowing them to get flyblown in the natural way. They were not a success though. I noticed at once that they passed through whole and undigested. Either the skins were too tough, or gentles themselves didn't suit him. Before giving them up, I squashed a few and tried those. He obviously loved them. So from then on, each gentle had to be pinched. It was a bit messy, but cheaper than the mealworms and quite as effective.

Snowy never really belonged to me. He treated me with great courtesy, but had obviously adopted my wife. He thought the world of her and displayed to her, whenever he saw her, with all the gallantry of his kind. That started, I think, because, as a fledgeling, he had to be fed so often – every hour at least. So I fed him about three times a day and she dealt with him every time he squawked between whiles. This, I may add, was no mean feat, since he didn't have to be very hungry to squawk. And, if he wasn't very hungry, he didn't want much and called out for more even sooner.

Right from the start we intended keeping Snowy indefinitely. It's pretty easy to get most hand-reared birds tame with their owners, but as they grow up, there is a real danger of them becoming as shy with everyone else as if they were wild. A spell in Children's Corner at our local zoo is as fine an antidote to this as anything I know. So many people file past continuously, that the shyest of animals or birds get so used to the sight of humans that they become completely blasé. I hadn't got anywhere to compare with that, so I built an aviary in the verandah, where there was a fairly constant procession of tradespeople. He was young enough not to care at the start, and he took people so much as a matter of course when he grew up that he never bothered.

There was nothing to worry about so far as feeding went, it was too easy. But as the weeks went on, our Snowy gradually acquired the colour of slush. He didn't pick up much grime from the floor, because I covered it every morning with fresh newspaper and burnt the soiled paper next day. This was partly so that any food he dropped should not be contaminated, and partly in self-defence. If it's necessary to keep a bird as large as a rook in the house, it's preferable to keep it clean. It didn't solve the problem of white feathers becoming greyer and greyer, though.

The obvious remedy was to provide a bath. Only then did I discover a remarkable gap in my knowledge. I had spent years reading everything that came my way about our native birds and beasts; I had devoured learned papers by scientists and pseudo-scientists on every subject from the migration of birds to the sexual habits of rabbits. Instead of memories of a misspent youth playing billiards, my mind harked back to nostalgic pictures of poachers and rat-catchers. Yet a lifetime spent in hoarding every scrap of knowledge about my wild neighbours had not even taught me whether rooks bathed in water, had dust baths or never took any baths at all.

I hung my head in shame.

Not daring to look any competent naturalist in the eye, I slunk out at dawn to find out for myself before my ignorance should become public. But I didn't find out. I saw rooks feeding their young and probing the fields for food. I saw rooks flying to their rookeries and preening their feathers in satisfaction at a job well done. I even saw them marching round the edge of a pond and drinking. But I could neither find a rook having a dust bath nor indulging in more conventional ablutions. For several days I would not admit defeat.

At last I made an excuse to visit a local landowner, who was proud to have several rookeries on his estate. I muddied the trail as best I could by talking for a while of anything but rooks. Then, all innocence, I turned the conversation to parasites of birds. 'Each pair of rooks,' I said, 'rear three or four young a year. You shoot comparatively few and yet the countryside does not become overrun with them. Something must kill a lot; what do you reckon it is?'

After a long and learned discussion, it became obvious that neither of us knew. The way was open for the next thrust. 'Parasites,' I said, 'could it be that? How do they get rid of their parasites?' The obvious remark trickled into the awkward silence. 'Lousy as a rook' – I don't suppose they do. Rooks are always said to be lousy birds. My way was open to go into the attack without loss of face. It soon became apparent that neither of us knew if or how rooks kept clean.

Next time, of course, it was easier. I mentioned to a nationally famous naturalist that neither I nor another friend, who should have known too, were certain about the hygiene of rooks. It was obvious, from the answer, that that made three of us. Feeling slightly less ignorant, I tried the same question on several of my friends, who could be expected to know, only to get the same result. Apparently it was one of those odd little gaps in our knowledge which are so surprising because they are so obvious.

I decided to find out for myself. Snowy's aviary was about

eight feet long and three or four wide. At one end I put a seed box of dry sandy soil and at the other a shallow dish of water.

As always, he took to all innovations with grave suspicion, and for some time, acted as if his pen was still empty. While I was waiting for the answer, I formulated my own theories. The jackdaws of my boyhood days had loved bathing in water. So had the wild ones and jays and magpies that I had watched. Logically, then, I expected my rook to take to his water pan and, sure enough, when the novelty wore off, he began to enjoy a daily bath with all the relish of a starling in a puddle. The grey of his feathers melted into a scintillating purity, which would not have disgraced a Christmas card. Once more, I savoured the thrill of making an original discovery, however unimportant. The fact that thousands of other people must have known the same thing didn't matter to me, because I hadn't been able to find anyone who did.

While he was young, up to about October of his first year, he had a bath almost every day, and often two if the weather was hot. For a long time he was as shy about it as a spinster with a poor figure, and, as the weather got colder, he became less and less keen until he wouldn't bother unless we gave him slightly warmed water. As spring came round, he started again, though he has never been quite so regular as during that first season.

I don't know how to tell a cock rook from a hen for certain, but I am almost positive that Snowy was a male. As the winter grew into spring, his cries for food became less strident and constant. I had taken an immense amount of trouble to persuade him to feed himself. Nothing but near-starvation had been a strong enough incentive. The one sight that sent him almost into a frenzy was the rich yellow yolk of a hard-boiled duck egg. So we crumbled this up on his clean newspaper before his morning feed. He hopped down and pecked at it and was disgusted to discover he was an utter failure. Sometimes his beak would snap harmlessly an inch above and, at others, he would hit the paper with beak still agape. There simply was no co-ordination between his beak and his eye. So we tried him with mealworms.

These are like the larvae of leather-jackets, so beloved by wild rooks, and they can crawl about as fast as the caterpillar of a cabbage-white. Snowy would pounce down and give one futile stab after another. We used to lay bets on the result and, at first, it was safe to lay slight odds on the mealworm. Before old Snowy had gauged his target correctly, there was usually time for it to gain the cover of a crack in the bricks.

We didn't make him feed himself entirely for many months. It's fatally easy to get a young bird or animal tame, by spending a great deal of trouble, and then, when the novelty wears off, all the good work is undone because a surge of independence recalls the instinct of the wild. So my wife went on feeding him by hand and, for months, when he felt sentimental, he would open his beak, quiver his wings and squawk like a fledgeling for dainties to be thrust down his maw.

It is not uncommon for birds reared under these con-ditions to get a fixation for the person who reared them, often to the entire exclusion of their own kind. Sure enough, it soon became obvious that our rook thought the world of my wife. Whenever she went near his aviary, he would hop importantly up to the wire and the pouch under his lower beak would inflate as if he were bringing her an offering of food. Such is the deception of men, of course, that it was but empty show. He would stand there, pecking empty air time and again in a queer, stilted, bowing motion, and the feathers on his legs would fluff out, for all the world as if he were wearing plus-fours. All the time he would burble a throaty, confidential chatter of endearments, the feathers along the top of his head would rise and his tail would flip and fan in a rhythm of ecstasy. It was a most moving and exciting display, showing such obvious affection that the stoniest heart could not remain unmoved.

To a very much smaller degree he would display to me, too. If I talked softly to him, calling him Joe in a deep voice – which he much preferred to Snowy – he would burble back to me and peck the air like a conjurer catching billiard balls. But it was obviously a platonic friendship and he would leave me, like a burnt-out flame, the moment he saw my wife.

By the time he was eight months old, we had bought our cottage. One of its attractions was the rookery in the garden and we were most anxious to see what would be the reaction of our albino; whether he would scold his wild relations or pine to be with them; whether they would acknowledge him in any way.

Before we could find that out, we had to solve the problem of moving him. Although he was so tame in his aviary he flew into a real panic if he happened to hop through the door out into the verandah. He would then flap madly round, bashing into the netting, the windows, or even the door, in a wild effort to get back 'home'. When he did regain security, he would sit on his perch with beak agape panting and terrified.

Obviously he was going to be a difficult customer to move without the risk of undoing our months of careful work. Opposite the back door of the cottage was an old room that had been a brew-house and bake-house combined. It was about ten feet long by eight feet wide and eight feet high. It made a very good indoor aviary. Outside, opening from the door of this room, I fitted up a large flight, about fifteen feet long and ten feet high, which I fitted with all the paraphernalia of swinging perches and branches that I thought would amuse him.

When we were ready to move, he was my only anxiety. So I fell back on an old falconer's trick. I decided to keep him in the dark until he got used to his new quarters.

It worked like a charm. The night we moved, I caught him and held him gently in my hands, while my wife drove the dozen miles or so to our new house. I had already blacked out the window of his room, so I went quietly in, popped him on a perch and shut the door behind me, leaving him in inky blackness. Next day, we crept in and just lifted a corner of the curtain long enough to stuff him with food, before leaving him quiet again. The day after that, we lifted just enough curtain to give subdued gloom and, gradually, we gauged the light by his ability not to panic, until he was free in his new quarters about a week later. The only trouble we had after that was with the doorway between his room and his flight. He would fly or hop through it, but it was quite impossible to carry him through. Rooks never naturally go into buildings, and this odd phobia about going through doors was his one concession to the wild.

As soon as he began to use his flight, we watched to see what his reaction would be to wild rooks. They simply might not have been there: there was just no more to it than that. When they flew over he never looked up. When they cawed he was apparently stone deaf. That did worry us a little. It is not uncommon for pure white animals and birds to suffer from deafness, and it hadn't occurred to us before that all the responses of our rook might have been stimulated visually. However, a few simple experiments put our minds at rest.

If his black brothers were beneath his notice, we wondered what he would do if we could procure another albino for company. I had had a good deal of correspondence with the veterinary surgeon from his native village, from whom I had discovered that there had been an average of one albino rook every year for six years. One had gone to London Zoo and died. One had had a leg amputated, due to an accident, but had not recovered. Two had been found dead after being blown out of the nest, and I saw several photographs of one reared by a gentleman who didn't answer any of the letters I wrote to him. I subsequently heard that that had died too. My Snowy seemed to be the sole survivor. I have since heard from a lady who had one also, though that was found several miles away and there is no certainty that it came from the Ashingdon rookery.

In the course of my correspondence with the vet, I discovered that the rookery in question was in the rectory garden and, since I had acquired my bird without any communication with the rector, I felt a little sheepish about it. My conscience is pretty elastic about poaching the odd rabbit, but few people can have poached an albino rook from the rector. Or, as in my case, 'received' one, which is worse. I wondered if he would ever find out.

Birds

My worst fears were realized when I got a letter next May from the rector's wife. She said she had heard that I had 'last year's albino' and, since 'this year's' specimen had been blown out of the nest and she was rearing it, she would be grateful to hear how mine had got on.

My thick skin served me well. I replied as if it were an everyday occurrence for folk to send me white rooks, poached from the rector. And I told her all I knew. The only thing that rather worried me, which was made plain in our subsequent correspondence, was that she had decided to keep her rook clipped. In point of fact, a member of the family clipped him out on one side, to prevent him escaping and coming to harm. I feared that if he escaped damage from cats, he would become subject to panics as he grew up and possibly damage himself by collisions caused by his lopsided flight.

As time went on, he did gradually develop a horror of lawn-mowers and certain strangers. It became more difficult to keep him at liberty and yet under control, and the rector's wife asked me if I would like to put him in with Snowy. They missed him very much; Snowy showed no enthusiasm; my wife and I were grateful and delighted.

We were extremely careful about the introduction. Snowy was on his own ground and in full possession of all his faculties. We feared for the safety of 'James', as the new one was called, who was exactly a year younger and deprived of the power of flight. So I built a cage within the aviary for James, and only let him out when one of us was there to see fair play.

As it turned out, it was a very good thing he couldn't fly. Right from the start, there was no doubt whatever that the newcomer was boss and he chivvied my Snowy whenever they met. But Snowy could fly up out of reach, so it was soon safe to leave them both together. The months crept on, and they both lived in the same aviary without ever getting on speaking terms.

We had been intrigued to watch the progress of Snowy's whiskers. At the base of the bill, young rooks have a bunch of short whiskers which are subsequently replaced by the familiar greyish-white skin which distinguishes adult rooks from crows in the distance.

It was a repetition of the 'bathing' controversy. Fine reputations were lost in arguments about it. Some said that the whiskers would moult off the first autumn and be replaced by the skin. They were wrong for a start, for he still had his nasal whiskers at a year old, when James came to join him. Others said that the whiskers didn't moult off but were worn away by the rooks' constant probing in turf for insects. So I fitted up one end of the aviary with a patch of thick turf, which I replaced with fresh as soon as it wilted. Both birds loved it, and pottered and poked about there by the hour for worms and insects, which were hiding in the roots. It served the invaluable purpose of keeping them occupied and free of the boredom so fatal to all captive creatures. They worked there for hours every day. But they did not wear off their whiskers.

Snowy lost his first. They came off with his moult (at about fifteen months old) and were never replaced. James lost his next autumn.

Time, too, healed their enmity. James remained perfectly friendly but never so sentimentally attached as Snowy. Perhaps this was because we didn't rear James and his allegiance was still with the rector's wife. But perhaps 'James' wasn't a cock bird after all. The way Snowy submitted to being chivvied, right at the start, reminded me of more than one woman-licked husband I have seen. And they chortled and chuckled to each other incessantly. Snowy learned to say 'Joe' to my wife and still displayed to her with all his old ardour. But as soon as she disappeared, he cast his eye at James.

Sadly, as so often happens with 'freaks' in the wild, both the rooks were delicate. Although the covered part of their aviary was light and warm, they spent most of their time in the open flight.

Before we could be certain if they were a true pair, Snowy contracted pneumonia and died. James lived on for a few months and suddenly showed symptoms of avian TB to which he succumbed almost at once.

CHAPTER SEVEN

THE POLECATS

One friendship, which I value very much, was made more in spite of my love of natural history than because of it. As a child, I had a longing for a stag beetle which I have never yet satisfied. Then, as I grew older and kept ferrets, I craved for a polecat, to compare the behaviour of very similar wild and domestic animals.

Soon after the war, I noticed two polecats in the Children's Corner of Dudley Zoo. I peered in their cage to discover, at close quarters, what external differences I could see between them and domestic ferrets. They looked to me exactly like very dark fitchets, but they voiced their opinion of mankind by spitting and hissing at anyone who came too close.

Soon after I had seen them, I innocently rang up the manager, and told him that I had kept ferrets all my life, and was hoping to get a pure polecat of my own. In the meantime, I should be grateful if he would allow me to mate one of my ferrets to one of his polecats, in the hope of producing a hybrid. He was obviously very busy and only said 'Yes, yes, bring it along some time,' as the least line of resistance.

Three days before Easter, one of my jill ferrets was well in season – it is a singularly obvious condition with ferrets – so I presented myself at the office, where I was introduced to the manager, Mr Risdon.

It was immediately obvious that he had forgotten all about our telephone conversation and was not anxious to be reminded. I stuck to my guns and, to afford temporary relief, he repeated the invitation to 'bring the ferret along some time'.

That was a tactical blunder of the first magnitude. I always have my jackets made with hare pockets, so I produced her on the spot from her concealment about my person. There was an awkward pause before I was told that we must first find the Head Keeper. So we embarked on a high-speed tour of the grounds. It is an exceptionally hilly zoo, and we raced up hill and down dale, from the raven to the aquarium, from the sea-lions to the wolves. The whole time, my new-found friend was telling me how desperately busy he was getting ready for Easter, and that he was overdue at the office for an appointment.

I felt almost ashamed of my thick skin. But it is an asset I have cultivated over the years and it rarely lets me down.

A Weasel in my Meatsafe

It became obvious to me that it was more than a coincidence we had not met his Head Keeper, at about the time he realized I was impervious to gentle hints. So he suggested, quite bluntly, that I should go away and bring her back after the rush of Easter. Patiently, I explained she was on heat at this very moment, and that it might be too late to mate her if I took her away. So he told me that his bitch polecat might savage and kill her. I said I had several more ferrets at home. His last fling was that my ferret might bite his polecats, in self-defence of course, and that they were irreplaceable. So I produced some string from my pocket and offered to muzzle her.

Why we ever became friends after such an exhibition, I shall never discover. Donald Risdon says now, in chivalrous retrospect, that I was not quite so brutal as I make out. Anyhow, he allowed me to put my jill ferret with his pair of polecats and told me to fetch her in a month.

When we made the introduction, we bated our breath. But my ferret was received as an honoured guest. They followed her round the pen, sniffing like dogs, and making the odd chattering affection call of the ferret tribe, which sounds as if it all comes on the intake of breath. There was no chattering, and none of the agonized screams that seem inseparable from domestic ferrets when they mate. I marked my diary and went away.

On my return, I could see that my ferret had not been mated, because she was still on heat. A peculiarity of ferrets is that the jills often remain on heat for weeks if their natural functions are denied, and then, instead of returning to normal, a high percentage of them die. So I invariably mate all mine, even if I have no intention of keeping the progeny.

132

The Polecats

The keeper on duty told me to leave her a bit longer as, since I had been there before, the bitch polecat had died. Once he had got over the loss, perhaps the dog polecat would show more interest in my ferret. I thought polecats must be very different from ferrets, if a minor matter like matrimony gave them any such inhibitions. Anyhow, the remaining polecat looked remarkably small for a dog. I asked the keeper if he were quite sure it was a male, but he said it was too savage to get close enough for certainty. I plucked up my courage, grasped the polecat's tail in my hand, and had a look for myself. This indignity was rewarded by a stink, thick and powerful enough to tear apart. The keeper was far more surprised than I to discover his two polecats had both been bitches.

Once more I returned to the office and this time Donald and I got on famously. Within ten minutes, I had permission to bring a hob ferret to mate to his polecat – she was as likely as a ferret to die if not – and he said that, if there were any result, I could have one of the kittens. Four were born, six weeks later, and when they were about five weeks old, I went over to claim my due.

She was a little bitch, almost black in colour, with typical polecat markings on the head. At first sight, she looked exactly like a dark fitchet ferret, but, right from the time she was quite a young kitten, she would hiss, if anyone approached, and dive for cover. I was anxious to give her every possible advantage, so I put her, with some young ferrets, in a turf run I had used to keep game-fowl. Round the edge I had sunk two rows of bricks to keep the rats out, and they served to keep the ferrets in.

The polecat kitten was quicker and more agile than the ferrets, even allowing for the fact that she had evidently had a touch of rickets. She could shuffle quicker than they could run.

Because of the space and the fact that she had young ferrets to play with, I didn't handle her as much as I should, and she never became really tame. But she grew a fine glossy coat and was my pride and joy.

Next spring I mated her to a ferret. By the end of winter she would tolerate being handled under normal conditions, but I was certain she would have been utterly impossible to touch if she had been excited by the thrill of the chase. Once I had mated her, though, I took no chances of inciting her to eat her young. I left her severely alone. She reared six grand kittens but, while they were with her, she would fly at any intruder like a wild cat. Apart from that, there was little apparent difference from a ferret.

That autumn, while I was at work, my wife was sent for by an irate poultry-keeper, who said there was a ferret in his fowl run.

Automatically, she mumbled the ferret-keeper's formula, 'We haven't lost a ferret, but if we can catch yours for you, we'll keep it till the owner calls for it.' My poacher friends of boyhood had told such stories of havoc and damage as to convince me of the folly of admitting ownership of any lost ferret. In truth, we hadn't lost a ferret. But my cherished hybrid polecat was not there when I got home.

Forlornly I searched everywhere that night, and at dawn the next day. And the next day and the day after that. Five days later, another neighbour reported a 'ferret' in his wood-pile. Coming home from work I went and waited quietly there, sucking over my teeth, as if decoying a stoat with imitation rabbit's squeals. Two beady eyes shone at me from the depths. I calculated the possibility of turning the pile over, but reckoned I couldn't finish by dark. So I got some nest boxes, baited them heavily with food, and left them when dusk came.

Next morning I was there before it was light, but there was nothing in my boxes. The food hadn't even been touched. If ferrets escape, they usually keep on walking. The fact that they have been seen in a rabbit burrow or thick cover today, is no criterion of future movements. It is as if ferrets have been domesticated for so long that they have lost all sense of territory and all homing instinct. So, naturally, I expected my polecat to move on.

I was surprised to come home a couple of days later to be told that my neighbour had captured the wandering 'ferret' in his garden. Apparently he had popped a dust-bin lid over it and put a couple of bricks on top to keep it in place. Agog with excitement, I rushed across. But when I lifted the lid there was nothing there. Polecats are not confined so easily.

I searched round and discovered, behind a garage, a huge pile of sticks and boxes and all the conglomerated rubbish of years. It looked just the sort of place to harbour rats. Since a ferret had been reported for several days, I reckoned this was where it would be lying. And, since a domestic ferret would not have stayed in one place, I was in high hopes that it was my polecat which was living there.

I went home, had a meal and returned half an hour later, laden like a sportsman setting out for a fishing contest. I, too, had all the necessary impedimenta. True, I had no rod. My bait consisted of a fresh cock's head tied on a piece of string. I had a thick leather hedging glove, a bag, a box with some straw, two nets (for rats, not fishing) and a flashlight. Like all good fishermen, I had a bottle of beer or so, to while away the time.

Fishermen usually begin operations by throwing ground-bait over the stretch they propose to operate upon. Instead of maggots, I threw small snippets of intestine, cut from the fowl I had killed. I threw them in a line along the edge of the wood-pile, in such a way that, if my quarry did emerge, she would eat one after the other, gradually 'feeding' further and further from the security of cover. When all was ready, I took the top off a bottle, settled down on an old box, like a fisherman on a stool, and cast my fowl's head into the likeliest gap.

I was overlooked by the bedrooms of a row of houses, and I imagine that anyone at a window must have taken special precautions that their premises were secure that night. There was I, squatting in the gathering dusk, motionless amongst the flies, dustbins and rubbish of a small suburban garden. At intervals, I would either take a swig from my bottle or haul in my cock's head, bloody and dripping on its string, and cast it back into a gap with all the solemnity that only fishermen affect.

As twilight faded into dark, I caught two glimpses of my quarry. She nibbled my bait but refused to be drawn into the open from her cover. My one consolation was that I saw enough to be certain that, at least, it was my polecat and not some stranger's ferret.

Any local householder, looking out next morning, could have been excused for supposing that I had been there all night. I was so enthralled at the possibility of recapture, that I was awake and about again by first light. This time I was luckier. Patient manipulation of my bait, on a string, eventually enticed my polecat far enough into the open for me to cut off her retreat. Within seconds she was in my bag on the way back to her enclosure.

After that she became incredibly difficult to confine. I have noticed that it is easy to keep ferrets in a turf run, provided there are about nine inches of bricks sunk around the edge. The ferrets often dig holes more than nine inches deep, but they appear to have no idea at all of coming up again. My polecat was a very different proposition. She would dig down like the ferrets, but she would also come up the other side of the fence as easily as a rabbit. I sunk slates at right-angles to the bricks, so that any hole, which emerged on the outside, must start well towards the centre of the run. This is adequate to keep rats in or out, but only lasted a while with the polecat. And when she wasn't digging, she spent hours a day climbing the netting and probing and thrusting for weak spots.

The result was that she got out several times, but in every case she went back to her pile of rubbish. Sometimes she would be out for several days before I caught her, but she always based her activities from this lair that she knew. She obviously still had enough of the instinct of wild animals to prompt her to acquire a territory. I was never completely certain what she lived on, but twice she was scarred by what must have been rat bites, and I think she got her water from the troughs the fowls used. Ferrets that escape are often quite stupid over water, and I believe more escaped ferrets die from thirst than starvation. Anyhow, I never had to worry much about my polecat escaping, for, even when she had not been at large for over eleven months, she went straight back for her rubbish heap and never did any damage to domestic stock while waiting for me to catch her.

Not so her daughter who was by far the trickiest animal I have ever handled to keep within the confines of captivity. She was the result of mating a son of the old lady from Dudley back to his mother. Inbreeding does little harm if not over-done and if weaklings are weeded out with complete ruth-lessness. Certainly no one would have called my hybrid polecat effete. I say polecat, because, although she had an admixture of domestic blood, she was as much a wild animal as her mother. She too escaped twice, and each time she took up residence strategically, not so that it would be difficult to recapture her, but so that she could fend for herself with comparative ease. Unlike her mother, she did an immense amount of damage when she got away, though luckily to my stock and not my neighbours'.

By coincidence she usually got away in the spring. Or perhaps captivity irked her when she felt the urge to breed. In either case, I usually seemed to have sitting hens about at the time and she went from coop to coop on her systematic pillages. I shudder to think how many favourite game-fowl she killed.

Like her mother, too, she was perfectly tractable during the winter. I fed her on meat and fowls' heads, and I could pick her up and mess her about as if she were some tame old ferret. When she had kittens, I respected her ancestry and left her alone, lest she devoured them and, within a fortnight, she had hissed and spat at me as if we had never met.

In the early summer I was invited to show some of my animals on the Children's Television. John Vernon, the producer, came over to discuss what I should bring as artistes. Bill Brock, my badger, was an obvious choice, as he was always well behaved and good value. At that time my young polecat had a litter which had not yet opened their eyes. The whole ferret tribe are exceptionally slow to develop. They are naked and blind when they are born. They grow their coats at between two and three weeks old and do not open their eyes until about five weeks old, by which time they are feeding on solid food.

I calculated that by the date of the programme, my young polecats, grandchildren of the Dudley bitch, would just about be opening their eyes and be at the squirmy stage, which would appeal to children. After tea John and I went across the field to the ferret runs to reconnoitre. The bitch polecat came running across the turf and I picked her up. It was a case of the quickness of my hand deceiving John's eye, for I did not want to give any clue that there was the least possibility of her biting me. In those days TV producers boggled at showing things to the children which might be fierce. She hissed a little as I took her up, but I was holding her firmly round the neck with her forefeet resting between my first and second fingers. So there was nothing she could do about it and she made the best of a bad job by lying quite still.

No one could help being enthralled by the kittens. There were six of them, as sleek as moles, with ungainly great heads like bull terriers. We enthused over them, for a while, and fell to discussing a layout for the programme. I became so absorbed that I forgot, for an instant, the character of the mother which lay in my hand. She was so still and comfortable there, with no sign of struggle, that she deceived me into feeling false security. She was so good that I released my hold, to allow her to relax and lie more comfortably along the palm of my hand. It was a mistake I shall never repeat.

With the speed of a rattlesnake she struck. Her teeth met through the first joint of the index finger of my right hand.

One second John Vernon and I had been lost in the intricacies of entertaining the children, the next I was doubled up in agony. This polecat weighed over three pounds and her strength was out of proportion to her size. My knucklebone gagged her jaws apart. The normal thing, to make a ferret loose, is to pinch its tail or front foot. It then lets go to chatter and bite its aggressor. I pinched – hard.

This, I suppose, had taken five seconds and John had not yet caught up with the action. All he could see was me shuffling from one foot to the other, pulling the most frightful faces and grunting. It hurt too much even to swear.

I gave up pinching her foot and got my left thumb over the back of her skull and finger under the base of her tongue where her mouth joined her throat. I squeezed again.

I remembered the bull terriers I had kept, and the thoughts flashed through my mind of the times I had had to throttle them off when they had been fighting. I can believe now that one's whole life passes before one's eyes, in cavalcade, before one drowns. Certainly I had time to remember almost every battle that both my bull terriers ever had, and they had had plenty, while I squeezed that polecat's throat.

Gradually her eyes closed and her jaws began to ease slightly, ever so slightly. As she relaxed I could see, in a glimmer of hope, that after perhaps three seconds' more agony I could snatch my tortured finger away. At last she went limp and, as my finger came away, I lifted the pressure from her throat, lest it harm her. If the mother died, her kittens were too young to do without her. Whatever the cost, I must not choke her beyond recovery. But in the fraction of a second before I snatched my finger away, for fear of tearing it, she got another hold. And like a bull terrier fighting, she grabbed again and again, to satisfy herself she was on to something solid.

By this time John Vernon had realized why I was grunting and cutting such an extraordinary caper. There was, by now, so much blood about that he couldn't fail to grasp the situation. I think his look of horror has imprinted itself on my memory as vividly as the pain. He wrung his hands and hopped from one foot to the other, begging to be told what to do. I was far too absorbed to oblige.

At last, after what seemed eternity, I choked her off. It hadn't done her as much harm as me and, after panting and sucking for breath for a few moments, she recovered enough to spit defiance at us and take her young out of harm's way. My finger had to be seen to be believed. There were over thirty deep punctures in it and, even today, if I stretch the skin, their scars bear witness to the worst hiding I ever took in my life.

It is one thing rearing young animals. At first they are too weak to hurt and, as they grow in confidence, there is usually plenty of warning before they get too rough for safety, or become actively vicious.

It is quite another thing reclaiming an adult which has once become savage. Obviously it was impossible to get the old polecat safe enough to handle in time for the television broadcast. When she had bitten me, I had said not a word. I had been far too busy getting her off. But I would not risk being within two feet of a microphone, on Children's

Hour, with a polecat on my finger. So we made a virtue of necessity. I took a strong glove and told the children that mother polecats sometimes get rather angry if you handle their kittens. My care in picking her up, even with a hedging glove, must have convinced them that I meant what I said.

That didn't improve matters as far as I was concerned. I don't keep animals to be looked at in cages. I keep them because I like them and I like the sentiment to be mutual. If she was fond of me, she concealed it very effectively. My affection for her had been temporarily estranged.

The difficult thing was to pluck up the courage to put my hand within striking distance of those wicked jaws again. It demanded more guts than I possessed. So I fell back on an old trick I had learned when I kept ferrets. I am always prepared to tame the most vicious ferret within a fortnight.

One of the lesser-known characteristics of all the Mustelidae is an extreme fondness for water. I do not mean, by that, that the stoats, weasels and ferrets will swim like otters for the sheer love of it. But they are thirsty creatures and not only will they drink water, but they seem to enjoy thrusting their snouts into it, blowing bubbles and generally splashing about. My weasel, for example, did not empty his drinking vessel by the end of the day because its volume was greater than his capacity. But I had to change it daily, for he splashed about as much as a bird having a bath, and he played in it so much that it was always a grubby puddle by evening. Ferrets and stoats and polecats are similar.

So if I have a savage ferret, I only allow it access to water twice a day, when I am there. It has to quench its thirst in my presence and, when it is satisfied, I remove the drinking vessel until my next visit. In extreme cases, I increase the enjoyment by substituting milk for water.

This worked like a charm on my polecat. I removed her from her large turf run to a cage, where people were passing regularly, so that she became used once more to the presence of human company. I fed her liberally on raw meat, to make her lazy, and removed her drinking water.

Next morning I offered a bowl of milk. She lapped greedily. It took two or three minutes to quench her thirst.

While she was thus preoccupied, I picked her up three or four times, caressing her with gentle hands, in a show of affection that was little short of traitorous. At first I was unusually cautious, always grasping her neck after moving my hand up from behind. The first she knew, she was gently but firmly lifted from her milk. It is easy to imagine that this would have infuriated her and only consolidated her venom. I do not believe that she ever associated cause and effect. Her one thought was to get back to her milk. She didn't get round to reasoning why it was suddenly beyond her reach.

When she got used to this handling, I took the training a stage further. I allowed her to commence drinking before picking her up, by approaching from the front, where she could see me. She was much too busy to bother.

The time had come to try another ferreter's trick. An old poacher pal once told me that all ferrets loved spittle. That the most savage ferret would quieten at once if a blob was proffered, warm, on the palm of the hand.

This, again, demanded the courage of my convictions. My finger was still raw and mangled. To offer the palm of my hand was rather like turning the other cheek, a virtue foreign to my nature. There was, however, a slight difference.

A Weasel in my Meatsafe

The risk was less than may be obvious at first sight. The right and proper way to approach any small animal which may take advances in the wrong spirit, is to proffer the back of the hand with fist firmly clenched. This stretches the skin so tightly that nothing – unless considerably bigger than a ferret – can get a grip, if it does strike. The teeth merely slide off, inflicting no more than a scratch. The same applies to a tightly-stretched palm, though, psychologically, it is a more difficult offering.

Like ferrets, my polecat appeared to have an unnatural craving for human saliva. When she had drunk her fill of milk and would drink no more, she would still lap the spittle from my palm as if she was parched.

The result was all I had hoped for. Within two or three weeks I could pick her up again, as easily as if she had been a tame ferret. I returned her to her run, where her young were now as big as she was, and twice as playful. In the summer evenings we used to go over and watch them gambolling about, with the undulations of little rocking-horses so typical of the whole tribe. Even my badger played with me in the same way, although the old dear looked quite foolish bouncing from front legs to back, time after time, as if he lived in toy-town.

This was where my courage rather failed me. I had lost my careless abandon and always took careful aim, before shooting out a hand to pick up the old bitch polecat. I didn't let her realize what was coming to her. Although she was perfectly tame again I have never quite recovered my confidence.

Not all so-called polecats have similar natures. A man who had brought one back from Germany, was put in touch with me by the Curator of Mammals at London Zoo.

The Polecats

Polecats are not good exhibits at zoos. They tend to spend most of their time rolled up asleep and many people mistake them for fitchet ferrets when they do see them.

I was particularly anxious to compare this German polecat with my own. My sceptical nature has long doubted whether there are any pure polecats left wild in the British Isles. So many people lose ferrets out rabbiting, that I often wonder if many animals, reported as polecats, are not really ferrets which have returned to their natural state. My own, and a number of previous experiments, had shown that the affinity between polecats and ferrets was so close, that they would easily hybridize, and that the resulting kittens were fertile and would breed, instead of being sterile. It seemed likely, therefore, that most wild polecats had at least been contaminated by crossing with escaped ferrets some time in their history.

When the German bitch arrived, it was quite obvious that the army had fondled her a very great deal. She was completely tame and very affectionate. She was lighter, both in colour and weight, than my own and her face was long and sharp like a ferret's. So much so, that I quickly came to the conclusion that someone had been practising a leg-pull. She had been offered to the zoo, but they had never seen her, so I would make it quite plain that I am not suggesting that they had been taken in. But whoever had had her on the Continent, was certainly more trusting than I.

When I introduced her into the polecat run, all hell was let loose. They chased her round like a rat until, in self-defence, she let off a stink that was well-nigh visible. I stood ready to intervene if they should really get to grips but it was more of a war of nerves. Nobody ever got hurt but they chivvied her mercilessly from pillar to post. That often happens with ferrets when a stranger is introduced, but within an hour or so, all is usually forgotten. My German friend must have possessed unusual gifts. They never did get used to her. Since she was so much smaller and thinner, I provided her with a large box, which could only be entered by a small hole, through which she could pass but my polecats could not. She just skulked inside until they were safely asleep in their own box. Then she would venture out and go round and round the run in a pathetic search for a hole through which she could escape to peace from persecution.

Humanity forbade me to prolong her discomfort. When I had satisfied myself that she and my polecats had nothing in common, I gave her to a friend, where she had a far better home than I could provide.

CHAPTER EIGHT

FAILURES

Although I have had perhaps more than my share of luck, I've had plenty of failures too. In the early days, a great many of the pets I came by either died, because of my ignorance, or grew up wild and untrusting, because of my lack of experience.

I caught some of them deliberately, like the nest of young starlings I tried to rear on bread and milk. They grew weaker and weaker, their feathers became bedraggled and matted with diarrhoea; as they died, their cries became more and more plaintive because, quite simply, they were getting the wrong kind of food. In a panic, I tried to put them back in the nest I had robbed, but the old birds had deserted it and would not come back.

I was about fifteen at the time, and a whole lifetime of tragedy completely filled my mind. Facing realities, I retrieved them from their nest and did what I could for them. One by one the survivors also weakened and died.

For days I thought of nothing else, and they haunted my dreams at night. It was like the nausea that follows drunkenness. I swore a solemn oath that I would never take the risk of causing such suffering again.

But I cannot throw off my craving for wild things any easier than a drunkard can forgo his alcohol. So I do the next best thing. Never, since those starlings, have I deliberately robbed some young thing of its parents without calculating, soberly, that the odds for survival are in its favour.

That does not mean that I have not often attempted to rear some animal or bird which has been found abandoned, or rescued from attack, and would inevitably have died otherwise.

My terrier once dug out a nest of water-voles, and I managed to save one from destruction. But it was much too young, and I hadn't yet appreciated that warmth, to young things, is at least as vital as food. I've always been very sorry about this, for I believe that a tame water-vole would make a perfectly charming companion. Although they look rather like rats, that is where the similarity ends. In the wild state they are far more docile and mild-natured and less suspicious.

The easiest things I have ever kept in captivity are hedgehogs. And in one way, they have been amongst the most satisfactory. That is that if you get tired of them, they can be turned loose in the garden, in the knowledge that they will come to no harm and neither will the garden.

I wouldn't call them either intelligent or affectionate, though I've had several which would unroll and allow me to slide my hand under the soft fur of their bellies and pick them up without pricking me. They very soon develop an addiction for bread and milk, so they will come to the house for a saucerful even after they have been set free. But I never learned much from tame hedgehogs, not nearly as much as I did from watching the wild ones.

My habit of wandering quietly about before breakfast and after supper, was always bringing me in contact with them. You can't be still very long in summer, when the dew is on the ground, without hearing the characteristic snuffling grunts, which must surely be as responsible for the word 'hog' as are the piggy eyes and snout. The thing that always intrigues me is the speed with which they cover the ground. They shuffle along, looking weak and oddly badger-like from behind. Yet, like badgers, they have quite a powerful, springy action, if you watch them from the side.

The most important thing that I learned, as a lad, about hedgehogs was that it is vital never to let a dog become interested in them. The temptation is almost irresistible because hedgehogs have such an alluring taint that they provide a superb opportunity of seeing just how well a dog works.

I used to sneak out about four or half past on summer mornings while it was still possible to see the tracks in the dew on close-cropped turf. Once a dog had become addicted to the scent, it was possible to watch exactly how accurately he followed the trail left by his quarry.

Addiction was the operative word. All that was needed to get him hooked was to find just one hedgehog and introduce the dog to it. The first jab on his sensitive nose, as he investigated the prickles, produced a love-hate relationship that endured for life. Thereafter he hated hedgehogs implacably but loved tracking them to their lair and digging them out.

If left to himself, any terrier worth his salt will worry at his quarry until he kills it. Not only is this extremely cruel to the harmless hedgehog but to the dog as well. The prickles lacerate his mouth and any that are dislodged are liable to become fast in his throat and fester. As a fringe bonus, the fleas that proliferate on hedgehogs, because they cannot scratch between their spines, are transferred to the dog. And, from the dog, they invade the house although they are not partial to human blood and will not thrive on people.

Nobody told me this as a child, so I had to learn it the hard way and my dog a harder way still.

Since then I've always taken the trouble to find a hedgehog as early in a pup's life as possible, and to teach him that harming hedgehogs is as great a crime as chicken killing. Even so, the scent of waterhens and hedgehogs are among the most exciting canine aromas and plenty of trained gun dogs will leave the trail of pheasant if they cross the trail of either.

As a boy, born on the edge of a town, I had more opportunities of observing hedgehogs than almost any other mammal. Large urban gardens with shrubberies and holly hedges covering a thick blanket of leaves provide ideal cover. There are quantities of worms and beetles for them to feed on, and hospitable human hosts who put out bread and milk at night.

So I saw hedgehogs most dawns and dusks when I was mooching round the fields at home. When alarmed, they freeze in their tracks, ready to roll into a ball if their suspicions are confirmed.

I once watched a fox hunting one. He crossed its scent trail a hundred yards from where it was feeding and followed its line as surely as a bloodhound. When he came up to it, he prodded it with his nose and it immediately rolled into a tight ball. The fox knew precisely how to deal with this. He nudged it gently over until it was lying on its back, and then he sat down to wait.

It must take quite a lot of muscular effort to roll up so tightly and, when the danger seemed to have passed, the hedgehog relaxed his muscles, which just left a tiny chink in the spiny armour into which the fox inserted his pointed muzzle.

It was then just possible to grip a sliver of flesh and fur where the spines joined the hairy underside of the victim. Of course, it caused the hedgehog to tighten into a ball again, but this caused the fox no real inconvenience because the gap where he had thrust his snout was as defenceless as the chink in the armour of Achilles' heel. All he had to do was to wait until the muscles relaxed again – and then to consolidate his grip by thrusting a little further into the soft defenceless belly. Soon he started to shake, as a dog will shake a rag in play, but the hedgehog did not find the game amusing.

Agonized screams proclaimed that he could resist no longer and he submitted by unrolling so that his attacker finished off the grisly job and settled down to eat him.

The astonishing thing to me is that hedgehogs are still

more common in towns than in the countryside, or at least in large tracts of countryside.

Scientists have examined the stomach contents of several thousand hedgehogs provided by gamekeepers, only to confirm that the damage they do to game chicks and eggs is minimal and that there is no logical reason for keepers to hound them. Nor is there convincing evidence that they suck the teats of cows, although they certainly gather round cattle lying down to chew the cud. The incentive is probably the insects that gather to feed in the warmth of bovine bodies. But, in spite of the evidence that should prove their innocence, they are persecuted almost as senselessly as badgers.

Although they should run slighter risks from men in towns, the population of urban foxes is often so high that the odds against survival of hedgehogs would seem slimmer than in peaceful countryside.

Although they are so harmless, they must be exceedingly tough because neither keepers nor rampaging motorists nor cunning foxes have managed to wipe them out, either in town or in deep country.

I have always loved hares, whether natural and undisturbed or as quarry for my lurchers. It takes a very exceptional dog to catch a hare and a fair amount of luck on top of that. I would rather see a good 'long dog' close behind her hare, and watch them turning and twisting and gradually working towards the hedge than any other sport there is. And if the dog catches more than one in ten, I am highly satisfied because it isn't the kill that matters, it's the chase.

Even better than that, I love to watch an old hare

performing her toilet during one of those wonderful late June mornings, when the rain is fine and warm and soft and sensuous. I watched one this summer for more than half an hour. The rain was so tepid that it was difficult to believe one was getting wet at all. I was in a favourite attitude, sprawled against the post and rail fence at the edge of a field of turf. I really stayed still for so long because I'd been watching a green woodpecker probing for ants. Then, out of the corner of my eye, I noticed a three-parts grown leveret feeding towards me across the corner of the field. The wind was right, for the rain was drifting across, like warm hill mist, into my face. And by the time the hare was about twenty yards away from me, I could see she was as wet as I was.

She sat up and looked round for a bit and scratched the base of her skull with her enormous hind foot, more, I felt, to pass the time than because of any itch. Then she shook off a veritable shower of shimmering raindrops, scattering them back where they had come from. That began her toilet proper.

She was as leggy as a thoroughbred racehorse. She licked and polished, till the green grass shone under her belly when she was standing on tiptoe. She licked her forefeet and brushed her fur, like a fastidious cat. She drew her fore-paws over her ears, and her coat gradually became so perfectly smooth and close that the rain collected in globules and ran off it as if she had been a dainty little wild duck, preening on the edge of a pool.

The whole time this ritual was going on, she never relaxed her vigilance. Her great brown eyes and twitching nostrils aided her questing ears in warning her of attack. At last, I flicked two finger-nails gently together, to see how alert she really was. No human ear could have detected the sound five yards away, but she spattered tiny sods of turf with the vigour of her take-off.

I was once offered four leverets which had been laid bare in their form by a man scything nettles. I naturally jumped at the chance. They would most certainly have been deserted and died, after such disturbance, if they had been left.

The very fact that there were four filled me with foreboding. Within about a couple of days of birth, it is usual for the mother to put each in a separate form, to limit her losses if a fox, stoat, or bird of prey comes by. So I knew these leverets of mine must be very, very young. Things were complicated even further by the fact that I was due to spend the week-end a hundred miles away doing a broadcast, and my wife was coming with me.

We did the best we could. We got a tin of powdered baby food, a rubber hot-water bottle and the inevitable fountain-pen filler. Right from the start, it was practically hopeless. They seemed to have no idea at all of sucking.

We didn't worry at first, because it's quite common for young animals to refuse a change of diet – and mother! – for the first twelve hours or so. The snag with these leverets was that they were so young there was a real danger that their little bodies would become too weak to assimilate the food, before their great spirits were sufficiently subjugated by hunger to let them try.

We arrived at our hotel, only to find that the proprietors objected to dogs and were generally unpleasant and inhospitable. The dogs were to form part of the broadcast, so I was able to bludgeon permission for them. Had the old woman who masqueraded as Mine Hostess discovered that our bedroom was also shared by four hares, her lack of hospitality would certainly have known no bounds.

The whole week-end was entirely spoilt for my wife by the fear that these four courageous mites would not survive their captivity. I was busy with my broadcast, so that most of the responsibility and heartache fell on her.

Whenever I saw her, she was either refilling the bottles from the hot-water system, or warming their food and trying again. Each was a perfect replica of an adult hare. Large brown eyes were wide open and alert: coat was glossy and perfect. Only the ears did not seem quite so large in proportion as those of an adult.

Whenever I could, I slipped off to help feed them. We cupped each in turn in the palm of our hands, inserted the tip of the glass tube and squeezed out a globule of milk by pinching the rubber. We had no success at all. Next day, two of them took about two or three spots and the others nothing. By this time their eyes were a shade duller and their coats a trace dishevelled. At night our spirits rose a little. Obviously the mites were ravenous and they gulped the first few drops. But they were far too easily satisfied for my liking. We tried at intervals during the night with no success. Then pneumonia set in and the last was dead by morning.

I suppose, in a way, we ought not to have tried. It should have been obvious to us from the start that they were too young and weak, as hares are notoriously difficult to rear. But their very youth and helplessness made it impossible to destroy them in cold blood. The urge to save them was too strong. Our heads were overruled by our hearts.

The summer before that, I'd had a grey squirrel for a

while. We'd been out with the dogs and rifles trying to thin out the grey squirrel population just outside Wolverhampton, where they were pretty common. One drey looked particularly well preserved, so I climbed the tree to push it out. As it fell to the ground, three young squirrels leapt out and the dogs caught them before they could take refuge in the rabbit-holes and bracken. When I got down, I opened up the drey to see what sort of a nest they had made, and was very surprised to see a squirrel still cowering there. Strictly speaking, I believe, it is necessary to get rather special permission from the Ministry of Agriculture to keep a grey squirrel in captivity. But I don't expect they award many prizes to folk who keep rats, and my rat had lasted some time and achieved some fame without getting me locked up, so I decided to take another risk with this squirrel.

He was never a moment's trouble. We fitted the rubber teat from a doll's bottle to a glass tube and suckled him from that. Like my rat, his cage was on the verandah, where he could see as many people as possible, and the larger he grew, the tamer he became. We got him weaned, first on to bread and milk and then on to nuts and wheat and toast and fruit. I used to prefer watching him eat wheat. Small as it was, he'd sit there twirling each grain between his forefeet as accurately and as fast as if it had been a hazel-nut.

Then, when he was three parts grown, he learned to leap and we began to get the real thrill and joy that squirrels provide. Grey squirrels, of course, are no better than rats and I never have the least hesitation about destroying them. But I was very fond of my tame rat and saw no reason why this squirrel should not be quite as attractive.

Failures

At first he would jump no more than a foot from his shelf on to my arm. But he needed no food or other inducement to try. It obviously came naturally to him and he loved it. So I nailed a labyrinth of bean sticks to the roof of the verandah and we started leaving him the run of the whole place when once the tradesfolk had been.

Monkeys, swinging on their trees at the zoo, were pale amateurs beside him. His tiny claws could get a grip where no grip seemed possible. He would career around his 'gymnasium' like the champion skater at an ice show. Just as fast, it seemed, with no more effort.

When we went in to play with him, we just became two more trees. He would whisk from one of us to the other, on to his shelf, round the roof and back to us before we could locate him. When he was tired, he'd retire to his box and utter queer guttural grunts of affection, until we thrust a finger in to scratch his belly, ears or chin. Never, after the first few days, did I see a flea on him, though the drey in which we found him was literally crawling. He was a delightful, affectionate character.

Then, when I went down one morning, he didn't come out of his box. I opened it to find he was curled up there, warm as if in sleep. It was a sleep from which he would never awake. Still clinging to his nose was the froth and bubble of arterial blood. The fact that he was still warm, was proof that he had been out that morning, and had not died in the cold small hours, which claim such a high proportion both of animals and people who die naturally. The first thing that occurred to me was that something had fallen on him and ruptured a lung. I could find nothing out of place. Perhaps he fell during his aerobatics and hurt himself internally. He was so friendly and agile and cheeky that it was easy to forget what a fragile scrap he was. The bars and trapezes of his little gymnasium were a mere eight feet from the floor and I'd always imagined wild squirrels could fall, with impunity, from the treetops if they missed their footing. We never found out what the cause was. But it was one of our minor tragedies that he should die so young.

We have not regretted all our failures, though. The household was swelled, for a while, by a young vixen and although she was tame and affectionate, we were never satisfied that she was really happy. Foxes are such wild things that I believe their freedom means more to them than anything in the world.

There has been an incredible change in the habits of foxes since before the last war. When I was a child, one of my greatest thrills was to get even the most fleeting glimpse of a fox. I would go out cub-hunting on foot in the early morning, or my father, if he had a farmer patient to visit after dark, would sometimes take me along to open the gates. Then, once in a while, a golden-red shadow would drift across the headlamp beam.

I shall never forget those cub-hunting days. I would sneak off round to the far side of the wood and lie up in some thick clump of rhododendrons. If the huntsman spotted me, he would sometimes winkle me out and send me back, lest I head the fox. He didn't want to kill too many cubs in those days, for they were none too plentiful. He merely wanted to disperse them over a wider area. But he didn't often catch me. I used to crouch there, silent and still, listening to the crackle of twigs from the hounds, and the crackling of hunting crops and clatter of horses.

Failures

I watched the first jays come stealing out ahead of the invasion, silent until they thought they were safe, and then screaming a warning to everything wild that the hounds were here to pillage and kill. I saw magpies and blackbirds, pigeons and pheasants, all with the safety of their wings to rely on. I saw rabbits, bewildered before they went to ground, and hares, stoats and rats all safe because, for the moment, they weren't the object of pursuit.

And then, sliding silently along, I often saw a fox. Sometimes it was an old one, which had seen this game before, and knew that safety lay in breaking out into the open and stealing off elsewhere. I love to see any dogs work, whether lurchers, hounds, or gun dogs, and my tummy would turn over with excitement as the pack hit the line and broke out into a chorus of screaming ecstasy.

But sometimes the fox I saw was a bewildered cub, whose world ended at the boundary of the wood. If he broke cover, he was lost and must run for the sake of running, until his lungs were bursting and the pack closed over him to tear him apart from his agony. So usually he turned back into the wood to try his luck at evading his enemies at close quarters, where he knew every inch of the ground. I used to go quite tense and live every yard of the drama with the quarry and not the hunters. If luck held out, the place so reeked of men, hounds and foxes that everywhere smelt alike. Then it was sometimes possible for a cub to cower under a tree-trunk or pile of rubbish, or in a ditch, and lie undetected until the trouble had blown over. And I would no more have split to the huntsman about any I had seen creep under cover, than he would have helped me get a cub to rear for myself.

By the end of the war, foxes were not only far more

plentiful but had spread to the very edges of town. At my old house, in Bloxwich, a fox was unheard of within a mile until then. Now, if you leave any poultry out, it is more likely to be 'foxed' than if it were deep in the country. The consequence is that a great many people, other than the Hunt, dig foxes out, and cubs are by no means hard to come by.

Having kept a good many British animals, it was therefore inevitable that I should try to keep a fox sooner or later. One April I was given a cub which had been dug out of some Forestry Commission land on the edge of Cannock Chase.

I reckoned she was between three and four weeks old, because she could only stand weakly, and her coat was still thick and grey and woolly.

She took to a bottle and baby food the morning after we got her and she was very little trouble the whole time she was on the bottle. Indeed, she was neither harder nor easier to rear than a domestic puppy of comparable age would be. As with Bill Brock, we did as much feeding as we could sitting by the fire, talking quite naturally, welcoming any friends who wanted to join the party.

By the time she was about eight weeks old, she was installed in the verandah and would come out of her box if anyone called 'Vixie'.

Dinah was her greatest joy. She is my dainty little whippet lurcher, sudden death to rats, stoats, or rabbits, but about the gentlest soul who ever drew breath with anything else. Vixie was just her mark. She was soft and furry and playful, and the hearthrug was a shambles every evening until the cub was tired enough to stop playing. Then she would climb on to my lap, curl up and go to sleep. As she grew, she extended her territory to take in the interminably long Victorian hall, kitchen and scullery, until eventually she knew the whole house, from her verandah at the back to the sitting-room at the front.

By then the weather was getting warmer and the evenings longer. Vixie reckoned it was time to increase her domain to include the garden. It was a silly little box of a garden, about thirty yards long by eight yards wide. Down the centre was a thick privet hedge, dividing the drive from the lawn; one side was bounded by a wall, and the other by an oak fence. Stabling spanned the bottom and a pair of gates joined our house to the garden wall.

It was ideal for our purpose, though, because neither dogs nor fox could escape easily. As the cub grew, there did come a time when she could have jumped over the fence, but her natural fear of the unknown always prevented her.

My wife is passionately fond of gardening and the first thing to suffer was a bed of tulips. They went over like ninepins, until there was only a solitary one left standing. Vixie looked at my wife and, in sheer bravado, bit off this sole survivor. It was a foolhardy act.

In vain, I repeated my story about all the neighbours having flowers but no one else had the grace and witchery of a cub in full sport. Once more we were banished to the field across the road.

This wasn't nearly as bad as it might seem. I tied a ferret line to the fox's collar and, when I wanted her, I only had to get within five yards and put a foot on the line, if she didn't come when called – and she rarely did.

I tried at this time to train her to a collar and lead,

because I remembered, in the old days, a bearded old rat-catcher from a village a few miles away, who never went out without his fox on a lead.

As I expected, she either rushed off, to be brought up short when she came to the end of the line, or she cowered down on her belly and stoutly refused to move. Every pup I ever had did the same thing, but it had been nothing but a battle of wits to cure them. My patience always outlasted theirs until they submitted to being led about in docility.

Night after night I tried it on Vixie. My wife comforted me by assuring me I had more patience than anyone she knew. But it is a wifely duty to offer sympathy and she didn't yet know Vixie. I would stay stock-still, waiting for her to move. When she did take it into her head to dart off, like a freshly hooked salmon, I 'played' her like a fisherman with a rod and line. Time went by and I discovered that all I was doing was alienating her affections. I admitted defeat and returned to the garden.

She reacted like a pampered woman. She fawned on me and flirted, until I really believed she meant it. Butter, it seemed, wouldn't melt in her mouth.

My wife had watched our battles with the lead and felt sorry for us both. It is quite out of character, on the one hand, for a fox to be led about like a dog. On the other hand, I take pride in my persistence and it went against the grain for me to admit defeat. So we returned to more natural games in the freedom of the garden.

This time Bingh Singhs, our Siamese cat, joined in. He suffered from incurable curiosity, and it was more than he could bear to watch Dinah and the cub playing catch-as-catch-can without joining in; and Muffit, the hunt terrier, tagged on at the end of the queue. Being firm believers in the adage about two being company, we limited the party. The cub and any one, but not more than one, of the others. Muffit took a bit of watching. She was a fiery little bitch, and we feared what might happen if the cub accidentally hurt her; she was bred to tackle foxes and I had dug quite a few out with her. So having proved the point that she was prepared to bury the hatchet and treat at least one fox as a friend, we limited the times they were together pretty drastically.

That left Dinah, Bingh and Vixie. As the cub grew up, we noticed her begin to be a little rougher with the cat. She would get up absolutely full speed and snap at his long tail in passing, to be yards away before the poor chap could so much as unsheath his claws. These snaps got harder and less playful.

This rather spiteful behaviour to the cat was the first of Vixie's wild instincts to be superimposed on the confidence she had gained through being reared artificially. Bingh Singhs joined Muffit on the 'forbidden' list.

Then, gradually, there came another expression in her eyes. A perpetual faraway look, imparting an almost ghostly appearance, as if her body were with us, loving and playing as of old, but her soul had returned to the woods. It was a horrible, uncanny sensation, and it was exaggerated because she began to have fits, like puppies do when teething.

One minute she would be tearing round the house or garden with Dinah, tails tucked up in the full enjoyment of a pair of boisterous kids, then suddenly she'd stop in her tracks, her eyes would glaze and she would immediately start rushing about again. But madly this time, not seeing where she was going. Sometimes she'd bang, full tilt, into a door or wall, or squeeze, for refuge, so tightly behind the gas-stove or under furniture, that she couldn't escape again without help when she was better. At last her tortured stampede would end in merciful oblivion. She would roll over stiff, frothing and unconscious.

There was nothing we could do to help. Naturally we hoped it was but a phase that would pass with the acquisition of her second teeth. Every fit she had undid just a little of the patient hours that had tamed her. She grew ever more nervous and jumpy, hating tradesfolk, strangers, indeed everyone but Dinah and ourselves.

Within two or three weeks her fits got more frequent until, finally, she never regained consciousness.

I did a post-mortem to see if she had been infected with worms, which often give puppies fits, but discovered that her liver was very enlarged and jaundiced. We mourned her passing in a way and swore never to keep another live fox from its liberty.

CHAPTER NINE

WEASELS

I once found a book lying about the house which was inscribed to the effect that it was a prize which some long-forgotten relative had won for doing well at school. There was no possible danger that I should ever emulate the feat, so I idly flicked the pages over to see what sort of stuff had been inflicted on this uncle to encourage him to continue being an example to his fellows.

The only thing that stuck in my mind was that it was a

translation of a collection of French short stories. Naturally they were of a type considered suitable for schoolboys and the only one that impressed me was about a woman who kept a tame weasel. This lady had a weasel confined in a bird-cage in her bedroom, and when she awoke she let the weasel out of its cage and it played on her bed until it was tired, when it went to ground in a hole in her eiderdown, and slept until breakfast-time.

I remember scoffing at the whole thing and thinking it was just the sort of stuff little prigs, who won prizes at school, would swallow. But, in spite of my scepticism, I couldn't help being intrigued. I longed for a weasel, to try for myself and I dreamed boyish day-dreams of using it as a ferret, to enliven wet afternoons catching mice in the house, and of having a pack of weasels which would hunt miniature quarry for me, like midget hounds.

But weasels are not easy to come by. I longed for one all my life; I searched for nests in the summer, and all my friends would ring me up if there was the least chance of digging out a litter, but I didn't have any luck.

As is often the way, my eventual source of supply was about the most unlikely possible person. He was Peter Harris, an Australian dirt-track rider. The fans, I gather, called him a Speedway Ace.

He had come over for a season's racing and was staying with Stacey Nash, who ran a threshing machine outfit at Wednesfield. Stacey and I knew each other of old. I'd had many an hour's sport ratting in the ricks he threshed, and we often used The Dog and Partridge, close to where he lived. So when he discovered Peter Harris was interested in the same things as I was, he brought him to the pub one night.

We got on fine, and wallowed in our mutual interests. I was simply overjoyed to find that he had a tame weasel.

'Teasey' he called it, and he took it to the Speedway as his mascot. He'd been helping with a rick that was being threshed, and this young weasel had been thrown up, with a sheaf of corn, on to the threshing-box. Quick as thought he'd popped it into his pocket and looked carefully round to find the rest of the litter. There wasn't any sign of the others, though, so he took the one he had back home to Stacey's and, by the time I saw it, it was quite the tamest wild animal I'd ever seen.

It was so tame that he could walk about anywhere he liked with it in his pocket and it would never try to escape. It would simply run up his jacket and down the other side into his other pocket. Every night it came in the house and, when I first saw it, it was even safe to be let loose in the garden and was easy to pick up again. If I hadn't seen that weasel for myself I simply would not have believed it. My respect for the man who had accomplished it knew no bounds, for I knew what such a feat must have meant in terms of patience.

The result was that Peter Harris and I became firm friends. He was intrigued with my animals and birds too, especially Bill Brock, who was as much a novelty to him as his weasel was to me. He began to tell me tales of life 'back home' and he said that out there, they decoy foxes with whistles and shoot them for their pelts. From anyone else I would have regarded that as a tall story, but I accepted it at face value from Peter. I asked him to teach me to do it too.

As a result I obtained an invitation from Mr Vernon of Hilton Park for us to try our luck on the foxes there. He was just as interested as I was, but he did make the proviso that we were welcome with a whistle, but please leave the gun at home!

So, when the corn was cut in September, my wife and I collected our Australian friend and presented ourselves at the Hall, where Mr Vernon joined us.

We set off about eight in the evening along the edge of some woods and Peter produced his decoy whistle from his pocket. It was a simple, round, flat, tin whistle, about as big as a florin, with a hole in each side. When it was blown in fairly long blasts, it produced a very good imitation of the squeals of agony of a rabbit in a trap. The theory was that no fox can resist the temptation of going to help a trapped rabbit out of its misery.

For an hour, the four of us crept round like poachers, except that wherever we went we tortured the cool of the evening with the most heart-rending screams. We didn't see a fox, but more than once we heard a jay start to chatter and call in the heart of the wood, and gradually come nearer, swearing and blaspheming as he came. I believe that he was plotting the course of a fox drawn by the cries, but we never got proof in the form of a view. The jay always ceased his tirade without ever coming into the open. The one odd, unlikely thing that happened, was that we went to a badger sett, where we were able to see a whole family of three-quarter-grown cubs playing with the sow. When we'd had our fill of watching them, we tried the effect of the whistle to see how quickly they disappeared. Nothing whatever happened. They never so much as stopped playing to see what it was. It might as well have been quite inaudible. But when I deliberately broke a twig, to prove to myself that they were not as deaf as they seemed, the whole lot vanished in a twinkling.

Out of the wood, past the badger sett, we came to the open park, which lay between the house and the monument. It was a bit of land I knew very well, for I'd been rabbiting and rook-shooting there all my life. The open park-land was spattered with little square spinneys, full of rabbits in the old days, and fair cover to hold the odd pheasant even then.

We stood with our backs to one of these little belts of trees and Peter had one last go with his whistle. Suddenly, from a clump about three hundred yards away, a tiny figure drifted into the open. It looked for all the world like a collie at a sheepdog trial. It came forward with the short, sharp rushes of a good dog working sheep. And after each little dash, it subsided on to its belly so low that the ground seemed almost to swallow it up.

The rabbit in Peter's tin whistle screamed in death agony. Never was a beast more tormented, never did a fox have prey more easily got for the fetching. Between each piteous burst, the silence was such that I could hear the breath of the others gasp in excitement, and my heart was thumping, so that I feared it would drive the quarry away.

But everything went according to plan. Each rush brought him fifteen or twenty yards nearer, never directly towards us but always by tortuous zigzags, as if he was driving invisible sheep to phantom pens. By now he was near enough to see clearly, and he proved to be no shaggy Welsh collie, but a glorious fox in the prime of life.

When he got within twenty yards, he realized that there was something odd about the rabbit he was stalking. He stopped dead in his tracks, lifted a querulous forepaw and tested the breeze suspiciously. In the instant before he whisked his brush and raced for cover and safety, we noticed another figure about ten yards from him and twenty from us. We'd been concentrating on the fox's approach so intently that we'd had eyes for nothing else. At first we thought it was another fox. Then, as it moved, we realized it was not a fox but a hare. The rabbit decoy had brought that up to see what was going on too. Perhaps it thought a leveret was in trouble. For an instant, both froze into immobility. Then they were gone.

When we got back, Peter lent me his fox-whistle and I made one like it, so that I have been able to recapture the thrill of watching wild foxes and hares come to my call. But I gather that the news caused quite a flutter in the local Hunt dovecote.

Within a few weeks, Peter had taken for himself an English bride and returned to Australia. The day before he sailed, he arrived at my house and left his beloved weasel, Teasey, in my care. I never had a present I valued more, though I couldn't help thinking it a little ironic that I should at last come by the animal that I'd wanted all my life, not from an English naturalist but from an Australian dirt-track rider.

At a guess, I should say that he was about seven months old, and Peter said he was never really vicious, even when he caught him. He'd started his captivity in the bottom of a dustbin from which he couldn't jump out. It sounds a bit grim, but it would be easy to reach slowly and gently over the edge to get him used to handling. But by the time I acquired him, he'd graduated to a wire-netting run with a little box to sleep in. I didn't want the run but was grateful for the box to start with, as it gave him a refuge he knew.

I put the whole box in a wire-netting run of my own and sat back to watch developments. His reactions were fascinating. For a few seconds all was still, and then his curiosity got the upper hand.

First a tiny chocolate nose appeared at the hole in his box, wrinkling and crinkling to test the air. When he found nothing wrong, his nose grew imperceptibly into a snake-like head which filled the hole. For a second it was perfectly still, with no indication that it was alive except the glittering eyes. Then with no apparent movement, the hole in the box was empty again. While I was still pondering over this little conjuring trick, the hole was suddenly full of weasel's head again, so motionless that it seemed quite impossible it had ever moved.

This odd little illusion was repeated, until it was impossible to forecast, with certainty, whether the hole in the box would be hidden by a weasel's head or empty.

Before there had been time to decide, the next act in the play was unfolding. With no apparent effort, the young weasel seemed to pour himself out of the hole in his box, down the side to touch the floor of the cage, and double like a hairpin back into the safety of his hole once more. It was a wonderful sinuous rhythm, giving him the appearance of the endless belt on a sewing machine, flowing smoothly out of his box to the floor and back without ever stopping. Only the limpid brilliance of his eyes betrayed the fact that the 'belt' was not continuous.

Imperceptibly, the tempo and space of the dance increased. Instead of returning immediately to his hole, he first went once round his box, then to one side of his cage. Time and time again he repeated each additional step, until he knew the whole routine as perfectly as a cabaret star. But he was infinitely more graceful, lithe and dynamic.

His purpose was different, too. He was not interested in the effect of his ritual on us, his audience. All that mattered to him was that it should become impossible to surprise him far from safety. Wherever he happened to be in the cage I had given him it might mean the difference between life and death if he could shoot through the hole in his sleeping-box in one automatic, unthinking dive.

I suppose we watched him for two hours. Never, for an instant, was he at rest. Round and round his cage he went, diving for cover time and time again, from every possible and impossible angle.

I'd often watched animals at play before, practising their escape routines, but never with such ruthless singleness of purpose as this. He was deadly serious and it was obvious that neither food nor sleep nor comfort were to be allowed to come between him and what security he could fashion for himself.

When he was satisfied that he was immune from physical surprise, his confidence swelled almost visibly. I opened the door of his cage and he began to take me into his orbit and include me in the rhythm. He reminded me of the sparrows of my childhood, which would pass through the same mesh of wire-netting round the fowl-pen so often that they eventually did it as a reflex action, instead of more slowly by conscious thought.

Self-preservation, the most important of all instincts, was only partly fulfilled by an escape routine. Simultaneously it was necessary to acquire a territory to be preserved as the hunting-ground, so necessary to all weasels.

Exactly the same technique was used. He repeated each

sortie over and over and over again, so often that recognition of his surroundings was completely automatic.

By the time he had settled down physically we knew each other intimately. He would clamber all over me and go to ground in my jacket pocket if he were tired or alarmed. His craving for affection was almost dog-like, and he would roll on his back by the hour while I tickled his belly or smoothed the silkiness of his ears. He would reciprocate by holding the tip of my finger in his forepaws and licking it, if he felt sentimental, or holding it like a puppy in his teeth, if his mood was more coquettish. I have never kept any wild animal which was so affectionate or unafraid. I didn't doubt a word of the story of the Frenchwoman who kept one in her bedroom.

A few weeks later, when we moved to our next house, we were poking about in odd corners to pass away a wet Sunday afternoon. The rain was bouncing off the window, the whole place was dank and derelict and full of the rubbish that builders bring, and we were both feeling rather depressed.

When we got to the cellar, we noticed an enormous old-fashioned meatsafe on the wall. It was a really magnificent affair, beautifully jointed and framed in oak, with double doors and a movable shelf. My wife was quite delighted. Obviously it was a relic of more spacious days, when our predecessors had real joints of meat that they could cut at and come again. We'd been married at the beginning of the war and had never got accustomed to such luxury. The prospect evaporated the rain and brightened the whole afternoon. She talked of the day it should be scrubbed and resplendent with new gauze, the pride of her larder.

For me, it held out quite different promise. I could visualize it neatly enamelled and fitted with wire-netting instead of perforated zinc, and instead of hunks of meat on its shelves, I could see my sprite of a weasel flickering around it like a golden leaf on the tongue of a whirlwind. The sky clouded over again on our argument, until we made a compromise. My wife settled for our first refrigerator and I was to have the old cupboard.

A week later the repairs had been completed and I had the weasel in my meatsafe.

Of course, it was an unlikely place to keep him, but he settled down as if that was where he had been born and bred. Once more, the first few hours were spent in perfecting a fresh escape routine, and the next few in memorizing his new territory. But once he'd acquired confidence again, the games he played with himself would have put a family of squirrels to shame. I had fitted him out with a new sleeping-box too, on the same principle as Bill Brock's. It was divided by a partition, with a hole through it, and one end had detachable bars, with which I could shut him in if I wanted to. So I was able to take him out with me, in his sleeping-box, without causing him the worry of being in fresh surroundings. And if he accidentally fell to the floor and took refuge behind heavy furniture, he was quite simple to catch again. All that was necessary was to put his box down close to him and he'd pop into it, because he regarded it as the safest sanctuary in the world.

Every night, after supper, we brought this sleeping-box with him into the sitting-room. The reason that weasels seem fairly easy to tame is that they have particularly good sight and will not jump off an object they know, unless it is clear how they will get back again. Always excepting terror-stricken jumps of blind panic, of course. Naturally if they are frightened enough, they jump first and think afterwards.

We take infinite precautions to see that no animal we ever have is frightened, if we can help it. One good fright will undo weeks and weeks of patience, when we've been trying to gain confidence. So Teasey used to come in the sitting-room at night and would play for an hour or so on the bookshelf without ever jumping off, for the simple reason that he would land in strange territory if he did, and might not be able to get back. In the same way I could put him on my jacket, if I were standing up, and he would climb all over it, explore my pockets, jump on my head but never jump down because he trusted me.

When he was obviously used to the sounds and smells of the sitting-room, we put his box on one of the easy chairs. At once he started the exploration routine that he'd done in the meatsafe, when establishing his territory. In, out and round, in his weird rhythmic pattern, until he knew every inch of the chair and could dive for the cover of his box like an electronic automaton. Then he got bolder and explored the carpet round the chair, extending his sorties until he reached the next chair and the settee and the desk.

By this time his domain was quite extensive, and it became impossible to reach the cover of his box, without crossing what he obviously considered to be dangerously open spaces. So he began to establish additional refuges, where he could dive for cover, at strategic points all over the room. One was behind a loose cushion on the settee; another under a very low armchair; one behind the fire-screen, one under the bookshelf and one, a delightfully long tunnel, behind the curtains on the window seat. If we rearranged the furniture or cushions he would find a new hideout at once, but he would still spend hour after hour dashing from one vantage point to the next to perfect his technique. My wife and I sat spellbound, watching the jigsaw of his actions building into a complicated pattern, which would be well nigh impossible to observe under natural conditions.

Then, when the work of assuring the safety of his escape was over, he began his play. From whichever retreat hid him for the moment, a wedge-shaped head and wicked pair of eyes would appear. Then out he'd roll, turning cartwheel after cartwheel, like an acrobat going round the circus ring. He moved so fast that it was impossible to distinguish where his head began and his tail finished. He was like a tiny inflated rubber tyre bowling round the room. Sometimes we thought this game was purely for exercise, since we could distinguish no pattern. In the same way my bitch, Dinah, tucks in her tail and weaves round the lawn, for the sheer joy of living and to convince herself that no pure-bred whippet can match her speed, to say nothing of other lurchers.

Sometimes the weasel used his dance as a cloak for attack, though. He usually chose me for his victim, and his cartwheeling twisted this way and that, over the carpet and up on to the settee beside me. The fabric had started life when we got married as uncut moquette, but Gremlin, the Siamese cat before Bingh Singhs, had cured that nonsense. He'd used one arm after another to pull the loose sheaths off his claws, until the moquette was well and truly cut and ragged. It was an ill wind that blew nobody any good. It made a perfect foothold for the weasel, who could run up and down the perpendicular arms of the furniture with the ease of a squirrel, because they were exactly rough enough to suit his claws.

When his gyrations fetched him up on the seat beside me, I always knew what the next act would be. Often I would be sitting, as I am at this moment, writing on a tray on my knee. The tiny scratching of the pen and the movements of my fingers were irresistible. From the cover of his dynamic camouflage, he could dive on to my hand, grasp my first finger in his forepaws with the strength of a tiny bear, and bite the fingertip with mock ferocity but, in reality, as gently as a kitten.

This mock fighting was one of his most endearing qualities. If I tickled his belly he'd roll on his back and attack as if his very life depended on it. Then he'd gradually relax, until he was licking the tips of my fingers, and croon his high-pitched little purring love song in a show of genuine affection. Whenever he heard my voice, if he were in his meatsafe, he'd come to the wire-netting and dance the rhythmic exercise pattern he'd evolved to amuse himself and me. If I took no notice, he'd fly into a rage and rattle the wire-netting with his fangs, until I melted and opened the door. Then, as I played with him, he'd purr his delight more musically than any cat and incredibly loud for his size.

When Peter Harris had given him to me, he'd said I was lucky to get him. Apparently he'd been playing in some grass a few nights earlier and Peter had thought he'd knocked himself out on a brick. 'He passed clean out,' he said, in his attractive Australian drawl, 'but I pulled his neck for him and he came round again.'

It had seemed a most extraordinary thing to happen to a weasel, for I'd never thought of one being clumsy enough to bash into anything hard enough to knock himself out, and I forgot all about it. Then, one night in the sitting-room, I suddenly noticed him on his side on the carpet, twitching a little but obviously 'out for the count'. I hadn't seen how it happened, but picked him up to have a look for any damage. He was still out and, as I held him, he came round and promptly bit me good and hard through the finger. It was obvious that he didn't know where he was for a moment and had acted subconsciously. So I put him away without solving the mystery of his accident.

A few night later, he did the same thing when I was actually watching him. He stopped dead in his tracks and rolled over twitching like a puppy in a teething fit.

As soon as it became obvious that violent exertion brought on these little fits, we had to restrict his movements. With mutual regret, I feel, he ceased to have the freedom of the room and I put him, instead, back on the top shelf of the bookcase. As far as the fits were concerned, the cure was immediate. He never had another outside his meatsafe. But the limitation of his freedom left him with surplus energy to work off. He had evolved a complicated exercise pattern in his safe, running from side to side, up the wire on to the shelf, down the back and repeating the pattern time and again at high speed. There is no question about this pacing of cages, which is so common in captive animals. It is not an attempt to escape. Once they are used to their cages and take official possession, as it were, of a territory, I believe they would be scared and worried elsewhere. But they pace the same pattern over and over again, in a form of almost subconscious exercise.

Teasey was so full of energy, when deprived of his evening gambols round the room, that his exercise got more and more violent. He drove himself relentlessly through his routine, faster and faster, until he developed a positive frenzy of exertion. The result, of course, was inevitable. He began to have fits in his cage.

Within a few weeks they became more and more frequent until I went down one morning to discover him dead.

I think we both missed him as much as if he'd been one of the dogs.

His death was a topic of interest among my naturalist friends. Weasels have two litters a year and they have few enemies prepared to prey on them, partly because of their ferocity and partly because of the powerful pong they give off when danger threatens. So it is reasonable to suppose that there would be a weasel population explosion if there were not some other form of population control.

Nature seems to have provided an exceptionally ruthless and effective answer. Both stoats and weasels are subject to attack by a minute nematode parasite *skryjabingylus*. For part of its life cycle, it is a tiny worm which enters the nostril and works its way, via the tear ducts, to the brain. Its arrival there causes a series of recurrent fits which gradually grow more and more severe until, at last, the victim fails to recover.

Miss Frances Pitt had a fine collection of mammal skulls and, when Teasey died, we examined her weasel skulls and found that a high proportion bore tell-tale perforations which would eventually have proved fatal if they had not met a premature death in the traps of keepers first.

This susceptibility to fits strengthened my theory that my stoats had not been 'foxing' when they seemed to feign death on being shut in a strange wire rat cage.

I don't believe that animals 'sham dead' in any deliberate attempt to outwit their captors. I believe that the sudden emotion makes them throw a fit, and that their unconsciousness often causes their captor to relax his attention. When they come round again, it must often be possible to steal safely away before they are noticed.

Shortly after Teasey died I was doing a broadcast in which it was possible to mention this loss and suggest to the listeners that if anyone had another weasel he didn't want, I should be very glad to find it a good home.

I had several amusing letters, including one from a man who had a weasel at the bottom of his garden, which he fondly imagined would yield to my blandishments, if I went and called it like Teasey, and another from a child who 'hadn't a weasel but I could have his ferret instead if I liked'. But nothing tangible developed for some months.

Then, one Sunday in the late summer, a child arrived at my father's house, grasping a fistful of limp grass in his grubby hand. The contents were not immediately obvious, but at last, from the centre of the grass, a blind, naked mite was produced, about half as big as a mouse. The children had been for a walk and had noticed a tiny fairy wailing in the grass, which turned out to be this young weasel.

How it came to be there will always be a mystery. Perhaps the bitch had been forced to move her litter to another nest in broad daylight. And as she crossed the open, she may have been so disturbed and frightened as to make her drop the kitten she carried in her mouth and bolt for safety. I don't think that a very plausible theory, because weasels are courageous mothers, not given to deserting their young easily. The only possibility that I think more likely is that she left the nest in a hurry, dragging one of her suckling young out as she came. I've often noticed that jill ferrets will come into their run with a young one grimly gripping a teat. And when that happens, they seem very slow to appreciate what all the whimpering is about.

For the moment, I was far more worried about what to do with the mite than where he came from. His mouth was far too small for an eye-dropper or fountain-pen filler; he would be almost impossible to keep warm without an incubator, as my normal tin and light bulb would keep the side he lay on all right but I feared the top half would be cold; he would be very difficult to keep clean. Animal mothers lick their progeny clean, including their excrement, and, if the young are too small to sponge, they get into a horrible mess when reared by humans. All in all, we were pretty depressed but took him home to see what we could do.

It so happened that a favourite and much petted ferret had had a litter the previous day. As soon as I remembered that I fetched her into the kitchen, determined that however unwilling a hostess she was, at least she should provide the nourishment. I held her gently but firmly in both hands, with her belly uppermost. One hand held her neck and shoulders, the other her hind legs. Between my hands was a double row of succulent teats.

My wife took the young weasel from his flannel and held him gently on the ferret's tummy. I don't know how long he had lain in the grass before the children had found him, but there was no doubt about the fact that he was hungry. Blindly he swung his head to and fro on his feeble neck until his questing lips stumbled across a nipple. At once he settled down to feed. He sucked and sucked for minutes on end, while I held the protesting ferret still, and my wife maintained his precarious balance.

We don't easily give in, but we realized that this young

weasel might not be weaned for another month. He would want feeding at short intervals in the night, as well as when I was away at work, and at least four hands were necessary. It was more than even we could face.

So I took a gamble. I rubbed him all over on the ferret's breast, squeezing tiny globules of milk out on to his coat. Then I took him to the nest of day-old ferret kittens, which were about his size, and I popped him in the centre of the squirming mass. He was easy to see, because he had the faintest trace of yellowish hair, to give his nakedness a slightly jaundiced look. The ferrets, on the other hand, were the purplish pink of new-born human babies. The gamble we took was that by the time the jill ferret got back, the young weasel would have acquired the same taint as her own young.

Within half an hour she was writhing and frantic to get to her babies. She rushed into the nest as soon as I let her go, took one sniff, as if deciding that her family had been contaminated, and began to lick back their polish quite indiscriminately. Next time I looked, the whole lot were lying warm, contented and well fed, including the weasel.

Within two days the young ferrets had grown half as much again as the young weasel, so that it was obvious he would get last pull at the milk supply. Weasels were nearly impossible to come by and ferrets were easy, so I disposed of all but one, leaving the old lady only one kitten and a foster child to care for. Even so the young ferret shoved off his bedmate and wolfed most of the food, so he joined his brother monopolists, leaving the weasel as sole survivor. I don't know if he was chilled before I got him, damaged when the children found him, or never settled to his change of diet. Whatever the cause, he grew longer and longer but never much thicker. His head gradually looked too big for his body and he became weaker and weaker, till at last, after nine days, he died.

The origin of my third weasel was quite as extraordinary, because an abnormal amount of luck is necessary to catch one alive and well.

My friend, Dennis Wintle, who is a naturalist specializing in waterfowl, was driving his car, when he noticed a bitch weasel take a string of five half-grown kittens across the road. Quick as a flash, he leapt from the car, pulled out his handkerchief and chased the litter, which scattered in all directions. They were not old enough to have learned the territory, and one popped down what proved to be a 'blind' hole by a fence post. Dennis dived at this like a rugby forward smothering a ball and managed to capture it in his handkerchief. Leaving his car where it was, he walked home, still holding it, and popped it into a box till I could arrive to collect it.

I had been trying actively to get a weasel for over twenty-five years, and I had spent hour upon hour watching them in their wild state. Yet it had never occurred to me that a young weasel, big enough to run about, would have been foolish enough to pop into a blind hole, and I should certainly have hesitated to pick one up with no more protection than a bit of linen. It was the opportunity of a lifetime, which ninety-nine people in a hundred would inevitably have missed.

Naturally I took him home in high glee, and we christened him Tick Weasel. He was as different from my old Teasey as chalk from cheese, and he has cured me once and for all of predicting what any animal will do judged solely from observation of others of his kind, especially captives.

He was well past the stage of needing milk and I started feeding him on raw meat right from the start. He took to fowl heads, which a local poultry dealer kindly supplies, but I had learned my lesson with Teasey, and gave variety. Teasey had been fed on raw butcher's meat when I had him, and was very difficult indeed to feed on other things. I had found that the stoats did very well on raw duck egg, so one day a week I moved all food from the meatsafe except one raw egg. As a result there is no difficulty at all with Tick. If one source of food dries up, he will do quite well on an alternative.

The immediate problem was to get him tame enough to handle before he grew strong enough to do real damage. For a week he bit me frightfully, but I persevered and gradually he came to realize that I didn't hurt him and that my hands were not the enemies his instinctive fear of the unknown had foretold. So as time went on, he bit me less often. The early bites, though shrewd enough to draw blood, did not really go deep. As he grew tamer and bit me less often, my fingers healed up and now I've hardly a scar to show.

Perhaps the most notable difference between these two weasels was the smell. Teasey had a faintly sweet stink, rather like a granary infested with mice, unless, of course, he was annoyed, when he let off musk. Tick has a powerful, pungent body odour. He lives in a cage rather smaller than the meatsafe, which hangs in the kitchen, so that he can see people moving about nearly all day. And sometimes, as I come in through the door, I get a breath-taking whiff of weasel, quite unlike anything I noticed with the first one. The floor of the pen is covered thickly with newspaper, which I change every night, and that scarcely smells at all.

But in his sleeping-box is an old sock, which he curls up in, and this has to be changed at least twice a week or it would fill the kitchen with its tang.

He is not nearly so affectionate as Teasey, who loved being handled and tickled. Tick hates to be picked up. He is a complete individualist who loves to clamber and race all over me, but I must never retaliate by grabbing him. Every evening I retire with him to the cloakroom, which is about the only room in our old house with no holes round the skirting. He comes out of his box, dives behind a pile of magazines to spy out the land, and then clambers up on to me for his game. He ricochets round me like the old one, down my sleeve, on to my hand and back round the other side. He never croons his affection, though he is always pleased to see me in a reserved, very masculine fashion, but for no apparent reason, he will suddenly stop in his tracks when he comes down my sleeve to the bare skin of my hand. As he stops he goes rigid, hisses quietly and lays hold like a ferret on a rabbit. There is no real malice about it, and as soon as I get him off, he plays again as if nothing had happened. But I shall be very glad when he grows out of it.

By contrast, he seemed to have odd ideas about learning to hunt. One night, as I went down the hall, I saw a mouse pop into the cloakroom. I shut the door and fetched Tick, to see how he took to hunting. The very moment he got in the room he knew there was a mouse there. He stiffened, the hairs on his tail came up like the bristles of a flue brush, and his one object was to jump off me on to the floor instead of the reverse.

When he did jump down, he seemed at a loss. He ran to the door and all over the room except behind the pile of magazines, which was the only possible cover for the mouse to hide. Initially it was obvious he was hunting by scent, because his quarry was motionless and silent, out of sight. Then, by chance he went behind the books and literally stumbled on his prey.

I don't know who was more surprised, but the mouse recovered first and ran off to hide behind the pedestal of the basin. Tick was very excited and set off in pursuit but didn't know where to look, and that particular corner was one he rarely visited. Next time they met up he gave a half-hearted bite, which did no more than accelerate the retreat. By this time the whole room stank to me of excited weasel and, I imagine, to him of frightened mouse. His nose was no further use, his quarry was out of sight, so he began to hunt by sound. Oddly enough his ears proved more accurate guides and also seemed to stimulate his natural instincts more than nose or eyes had done. He worked up into a frenzy of excitement, located his prey as if by radar, and killed it as clean as felling an ox, the next time they met. When I caught him to put him back in his cage he bit me good and properly too, for not waiting for him to calm down.

All the same, he is a captivating sprite and a real challenge both to my ingenuity and my fortitude. He was, I suppose, rather old when he came to me. But I am gradually winning his confidence and I hope, within a few more months, that we shall get on as well as Teasey and I did.

Looking back, I'm eternally grateful to that old French story, which affected me so profoundly that I spent the rest of my life in searching for a weasel to prove, for myself, whether or not it was true. In the years before I succeeded, I achieved terms of intimate friendship with a great many animals and birds I might never otherwise have met. And, quite as important to me, I have got to know a good many people with similar interests. If I have luck, there should be left to me about the same period in the future to extend and enrich both spheres of friendship.

E.cN